The Patrick Moore Practical Astronomy Series

More information about this series at http://www.springer.com/series/3192

The Vixen Star Book User Guide

How to Use the Star Book TEN
and the Original Star Book

James L. Chen
Adam Chen

 Springer

James L. Chen
Shenandoah Astronomical Society
Gore, VA, USA

Adam Chen
Baltimore, MD, USA

ISSN 1431-9756   ISSN 2197-6562   (electronic)
The Patrick Moore Practical Astronomy Series
ISBN 978-3-319-21592-1   ISBN 978-3-319-21593-8   (eBook)
DOI 10.1007/978-3-319-21593-8

Library of Congress Control Number: 2015946604

Springer Cham Heidelberg New York Dordrecht London

Printed on acid-free paper

Springer International Publishing AG Switzerland is part of Springer Science+Business Media (www.springer.com)

This book is dedicated to my friends Gary and Sherry Hand,
whose support for all of my books
has been invaluable.

Introduction

One evening, around 8 o'clock on a cold, clear winter's evening, I received a distressed phone call from a teenager who had received as an extraordinary Christmas gift a Vixen VC200L 8″ catadioptric telescope on a Vixen SXD equatorial mount with the Star Book GoTo computer system. She was having extreme difficulties getting the telescope to work.

My contact information had been given to her by the local telescope shop that had sold the telescope and telescope mount to her mother. I was (and still am) the shop's local expert on using the Star Book equipped mounts. The teenager on the phone lived in an affluent suburb of Washington, D.C., and I was over an hour-and-a-half away by car. I tried to solve her telescope problems over the telephone, since instructing her in person was inconvenient and would be a last resort.

The teenager's problem was, using the Star Book to search for a planet or star, the equatorial mount motors would whirr, and the gears would grind, and the telescope would then point directly to the ground! Obviously, something went wrong!

Gradually, I stepped the young lady through the setup process for the Star Book, making sure the longitude and latitude of her location had been properly entered into the Star Book's memory, along with the date and all the time data. Being assured that the polar axis of the mount was aimed northward, and that she was following the alignment procedures, I realized her problems were occurring during the initial alignment of the mount. The Star Book wasn't cooperating and continually pointed to the ground. In a flash, I recognized the problem. At the beginning of the alignment process, the telescope and mount needed to be aimed to the West as its "Home" position. She had been pointing the telescope East. She had misinterpreted the picture displayed on the Star Book screen and pointed the telescope to the East, a Home position that was 180° from the proper direction. After making that correction, and following my instructions, her telescope was up and running, finding galaxies, nebulae, and planets to her hearts content. I haven't heard from her since.

As for me, for years, I had my 130 mm f/8 apochromatic refractor mounted on an early 1950s Unitron equatorial mount equipped with a mechanical Polarex clock drive. The mount was heavy and bulky, and the clock drive, although a curiosity at star parties, was a maintenance hog, needing continual lubrication and subject to wear and breakage to the weight drive cable that powered the device. Although fun to watch in action, with the weight descending, rotating gimbals and spinning fly-wheels, the Unitron mount was ancient, cumbersome, and labor-intensive to use. On top of all that, it really didn't track that well. It is difficult for me to read the 1950s advertisements touting this mount as photographic ready. A review of my observing log books revealed that I had used this telescope and mount combination only twice in the previous 5 years! It was time for a change. With the help of friends in the amateur astronomy world, I gathered information and managed to get some hands-on time with several commercially available equatorial mounts that would handle the payload that my telescope presented. In the end, I chose the Vixen Sphinx SXW as the most portable and capable mount. I have used this mount for several years now, with my usage of my 130 mm apo and SXW combination expanding to at least once or twice a week! The Star Book GoTo system has been a joy to use, and invaluable in locating deep sky objects in an efficient and precise manner. When I used it, I would average three or four objects observed for the night with my old Unitron mount. With the Vixen SXW and Star Book, I now observe between 10 and 15 deep sky objects on an average night. According to my observing logs, I have actually peaked at 25 deep sky and planet observations during an exceptional evening!

It's stories like these that made me decide to write this book. As with computer GoTo telescopes made by other manufacturers, there is a learning curve involved in using the Vixen Star Book. As good as the Vixen manual is, users benefit from an experienced hand at the controls. This book's goal is to provide hints, tips, and aid in using the Star Book TEN and the original Star Book, and make the learning curve shorter and effortless.

James L. Chen
May, 2015

Acknowledgements

A big Thank You to the following people who made this book possible:

To Brian Deis, Mr. Star Guy and to Vixen Optics, for the loan of the Vixen SX2 and Star Book TEN and Star Book One. And for the access to Vixen official photo files and documentation. Your support was invaluable.

To Gary and Sherry Hand of Hands-On-Optics, for providing technical support, conceptual ideas, and encouragement.

To my wife Vickie for her encouragement, support, and her proofreading and critiquing skills.

To my son Adam for his graphics abilities and valuable photographic suggestions and contributions.

To my son Alex for serving as a soundboard for some of my ideas for the book, and making valuable suggestions and contributions.

And to Nora Rawn of Springer, who gave a fledgling first-time author a chance, for supporting my book concepts, and being a good audience for my jokes.

Contents

Chapter 1

A Brief History of Vixen Computerized Telescope Mounts for Amateurs

In the ultimate mating of two hobbies, computers and astronomy, computer controlled telescopes have captured the backyard astronomer's imagination and pocketbook. Known collectively as GoTo telescopes, this advanced technology is fascinating to watch the mount as it proceeds to point the telescope from object to object with precision, accompanied with the sounds of motors whirring and gears meshing.

A GoTo telescope mount is quite simply a telescope system that is able to find celestial objects in the night sky, and then track them. The GoTo mount can be set up in an alt-azimuth or equatorial fashion, and after the proper alignment procedure, the finderscope is no longer needed for the rest of the evening. Some of the newer GoTo telescopes have electronics and CCD cameras that will perform the alignment procedure automatically.

These telescope mounts are wonderful pieces of technology. The GoTo technology allows for more efficient use of observing time by quickly finding objects in the night sky. Built into the hand controller is a microprocessor, firmware, and built-in memory catalog of the positions of thousands of stars, galaxies, nebulae, open star clusters, globular clusters, planetary nebulae, our solar system planets, and the Moon. Complex algorithms developed and refined over years with improvements in encoders and motor technology have made the GoTo telescope an accepted and desirable telescope feature. Computer controlled telescopes can help it's owner to overcome the fear of looking ridiculous while others watch; no longer will the telescope owner appear incompetent as he tries to find celestial wonders—now he only looks ridiculous as he tries to remember how to set up his telescope!

There is an ongoing debate within the amateur astronomy community on the merits of computer guided and computer controlled telescopes. The hardcore

© Springer International Publishing Switzerland 2016
J.L. Chen, A. Chen, *The Vixen Star Book User Guide*, The Patrick Moore
Practical Astronomy Series, DOI 10.1007/978-3-319-21593-8_1

conservative backyard astronomers argue that a beginner or novice individual is better served learning the skies without electronic aids, as generations of stargazers have done. There is merit to this argument. However, in these days of increasing light pollution in urban and suburban neighborhoods, seeing landmark stars used for "starhopping" to locate deep sky objects is becoming increasingly difficult and frustrating to a backyard astronomer, particularly to the beginner or novice. A computerized GoTo telescope and mount that relies on alignment with bright first or second magnitude stars greatly eases the frustrations of the hobby. The search time for a celestial object can be reduced from tens of minutes to mere seconds! With the electronics aiding the observer in finding the deep sky objects, a suburban observer can then take advantage of modern filter technology in overcoming the light pollution in their area. Cheers to the miracle of nebula filters, light pollution filters, and color filters!

Of course, in the worst of urban environments, even using a GoTo telescope and mount can be challenging, especially if bright stars are impossible to see for alignment purposes or otherwise. For instance, in the middle of brightly lit Las Vegas, the only bright stars visible are Wayne Newton, Celine Dion, and a variety of Elvis impersonators!

The era of computerized GoTo telescopes began in 1984. Computer controlled telescopes took form during the same period as the development of personal computers. During the 1980s, the U.S. telescope company Celestron formed a business relationship with Vixen Company, Ltd of Japan. The American company featured its home grown Schmidt-Cassegrain telescope, while importing the Japanese refractors, eyepieces, and equatorial mounts from Vixen, and marketing them under the Celestron brand. Introduced in 1984, Celestron Compustar 14 was the first computer controlled telescope offered for the consumer. The Compustar 14 is a large and heavy catadioptric telescope, designed for permanent installation in an observatory. Concurrently, Vixen of Japan developed the Sky Sensor, an economical system consisting of a Go To computer control system with motors designed to attach onto their portable German equatorial mount known as the Super Polaris.

The landmark Sky Sensor system was remarkable for its time. As the first consumer affordable GoTo system, it had 472 nebulae, star clusters, and galaxies stored in its memory. This is small, as compared to today's GoTo systems that have 30,000, 40,000, or more stored in their databases.

The reader is cautioned to understand that database claims are sometimes inflated and not necessarily truthful. There are a number of multiple counts for a single object. For instance, the Andromeda Galaxy counts as one object; M31 is an additional object; NGC 224 as another object. Thus the same object is counted as three separate objects in some manufacturer's database claims.

In 1984 standards, the Sky Sensor was revolutionary. The Sky Sensor data base contained all the Messier objects, NGC objects brighter than 10th magnitude, and 285 stars brighter than 3.5 magnitude.

Installation of the Sky Sensor onto a Super Polaris mount required a little mechanical dexterity, but could handled by the end user. And if not, the local dealers were experienced in installing of the right ascension motor and electronics card,

Fig. 1.1 The Sky Sensor computer controller (Hands-on-Optics Used Equipment archives)

declination motor and electronics card, gear shafts and pressure plates, and clutch knobs. Plug in the Sky Sensor controller and power supply, and the system was ready for use.

The keyboard, as seen in Fig. 1.1, was a bit archaic. Note the use of CR for carriage return instead of an Enter key! The art of human factors engineering had not yet entered into the design of telescope control. The end user faced a bit of a learning curve in operating the Sky Sensor. The system was not as responsive, accurate, nor as quick as today's modern GoTo systems, but as a first generation device it showed the way to the future.

In 1992, Meade Instruments introduced the LX200 series of fork mounted Schmidt-Cassegrain telescopes (SCT). Early 8 and 10 in. models were produced contained software bugs and were unreliable telescopes. Over time Meade was able to refine the LX200 models to become a very capable platform, with the product line extending to larger models, of 12 and 16 in. sizes, telescopes more at home in a college or NASA observatory than in the backyard. In August 1996 Celestron countered with the Ultima 2000 series telescopes—but they delayed shipping until 1997 until the software bugs were worked out. The initial offering was an 8 in. SCT Ultima 2000, which was a lightweight, rigid, and easy to use telescope (Fig. 1.2).

Meanwhile in the late 1990s, Vixen issued a revised version of their GoTo system, named the Sky Sensor 3. The Sky Sensor 3 featured an updated hand controller and other hardware. The database was still the same size (Fig. 1.3).

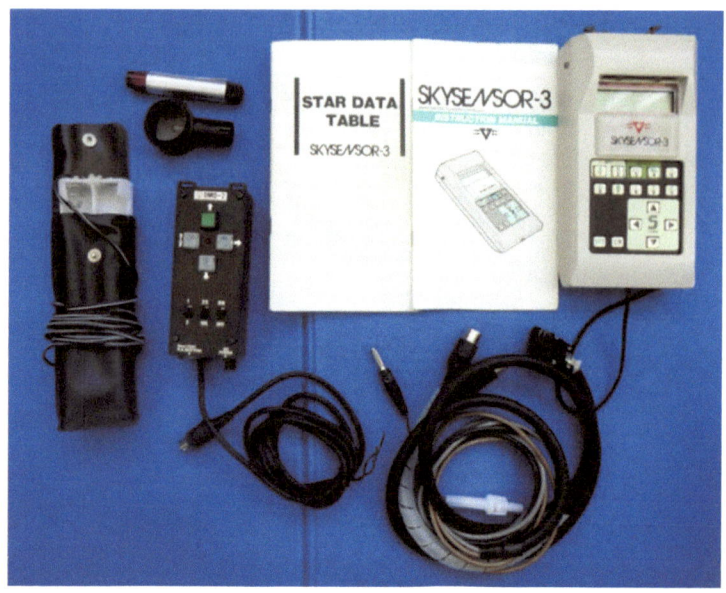

Fig. 1.2 The Vixen Sky Sensor 3 (Hands-on-Optics Used Equipment archives)

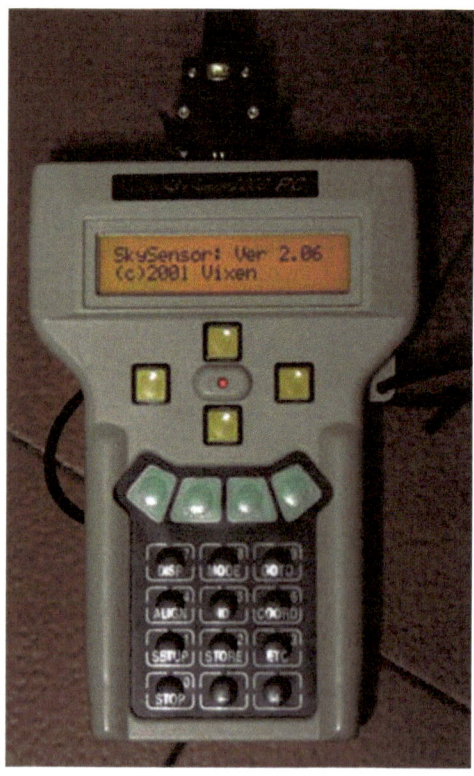

Fig. 1.3 The Sky Sensor 2000 series (Hands-on-Optics Used Equipment archives)

By 2000, Vixen introduced another revision to their venerable Sky Sensor series, now known as the Sky Sensor 2000. The SkySensor 2000 system was vastly refined and improved over the previous Sky Sensor models. The SkySensor 2000 could be retrofitted for use with the Vixen GP, GP-DX, GP-E, SP or SP-DX equatorial mounts to provide highly accurate "Go To" pointing and tracking of celestial objects in a vastly expanded data base that now included the planets, Moon, Sun, and thousands of deep sky objects from Messier, NGC, IC, UGC, SAO, GCVS catalogs, for a total of 13,942 celestial objects. The revised system simplified the initial setup and was easier to operate. The slewing rate was improved up to 1200× that of Sidereal rate (5–3/4 Deg. per second). The Sky Sensor 2000 incorporated the most accurate of the tracking control systems for the time by including Periodic Error Correction (PEC) circuitry to reduce the amplitude of worm gear periodic errors.

In the new millennia, major developments in GoTo telescope technology has been introduced into the consumer market. Meade and Celestron have introduced and refined their Autostar and Nexstar GoTo systems for fork mount and German mount designs. Databases of these telescope computer systems have been expanded to the 30,000–40,000 celestial objects range. Pointing precision and tracking accuracies have been greatly improved. The ease of setup has been improved. Many other manufacturers have joined the GoTo mount revolution, with offerings from Losmandy, Orion, Astro-Physics, Takahashi, iOptron, and many more. The computerized GoTo telescope mount has come of age.

Vixen did not stand on its laurels. The drawback of all previous GoTo telescope computer systems, including the Autostar, Nexstar, and even the early Vixen Sky Sensor systems, is the reliance on text as the mode of user interface and operation. Modern desktop and laptop personal computers have long since graduated to graphics-based operating systems, such as Microsoft Windows and Apple OS. Vixen recognized the trend in the personal computer world and applied graphics technology to their latest designs. In 2004, Vixen introduced the revolutionary Sphinx SXW equatorial mount, equipped with the first computer GoTo system to use a graphics user interface, the Star Book. By 2013, the major re-design and upgrade of the Vixen GoTo system was introduced, the Star Book Ten, with improved user-interface and a database exceeding 272,000 objects!.

The chapters that follow will cover in depth the Vixen mounts equipped with the Star Book Ten and original Star Book GoTo systems. An effort is made to clarify the sometimes confusing instruction manuals, while providing tips and hints on the Star Book TEN and Star Book use. The Vixen instructional manuals are well written, but often an alternative description of how things work is helpful to a newcomer to the Star Books.

Most importantly, the lessons learned from an experienced Star Book TEN and Star Book owner are offered in separate chapters for the Star Book TEN and the original Star Book.

There is also a chapter on Vixen's variations on the Star Book theme, the Star Book S and the Star Book One. As with all Vixen products, there are a myriad of accessories for Sphinx mounts that are also highlighted in a chapter within.

Included in the appendices are firmware update procedures for both the Star Book TEN and the original Star Book. Also included is an appendix is the procedure for loading into the Star Book firmware comet orbital elements. These procedures are available on Vixen's website, and are provided in this book as a courtesy. The reader is cautioned to throughly study the update procedures, reference and study the Vixen website for any changes before attempting any firmware updates or comet data download. Please note: the author recommends all updates to the Star Book TEN or Star Book be accomplished by the Vixen dealer or by Vixen itself.

Chapter 2

Vixen Sphinx Mounts
Using Star Book
Technology

It is important to get familiar with the family of Vixen telescope mounts with the integrated Star Book and Star Book TEN GoTo technology. In Vixen nomenclature, these equatorial mounts are their Sphinx line (Fig. 2.1).

Common to all of the Sphinx mounts, first- and second-generation, is the unique internal mounting of the right ascension (RA) and declination (DEC) motors. This internal location of the motors, electronics, and gears within the declination housing of the Sphinx mount acts as a built-in counterweight. Therefore, the amount of actual counterweights used to balance the mount with its telescope is less, resulting in an overall lighter and more portable equatorial mount. All wiring for the RA and DEC motors are also internal, achieving a clutter-free telescope setup.

The Vixen Sphinx mount system is designed in a modular fashion, allowing the user the flexibility for adapting the mount in a multiple of configurations. A more detailed discussion of the various configuration options will be addressed in a later chapter.

The original Vixen Sphinx mounts, with the original product introduction in 2004, included the original Sphinx SXW, the somewhat heavier duty SXD, and the king of the Sphinx line the ATLUX (Fig. 2.2).

The Sphinx SXW specifications, aside from the Star Book (more details later in the book), are quite respectable: 180-tooth worm gears driving it (implying smoothness and accuracy suitable for astrophotography), and the full-up weight of the equatorial head, tripod, and one counterweight is about 30 lb, with a rated capacity of 26.5 lb. Applying excellent system engineering principles to the SXW design, the Sphinx mount features a retractable counterweight shaft, fully enclosed servo motors placed inside the mount housing to provide counterbalance and minimize (but not eliminate!) the role for counterweights, and no external cables except to

© Springer International Publishing Switzerland 2016

J.L. Chen, A. Chen, *The Vixen Star Book User Guide*, The Patrick Moore Practical Astronomy Series, DOI 10.1007/978-3-319-21593-8_2

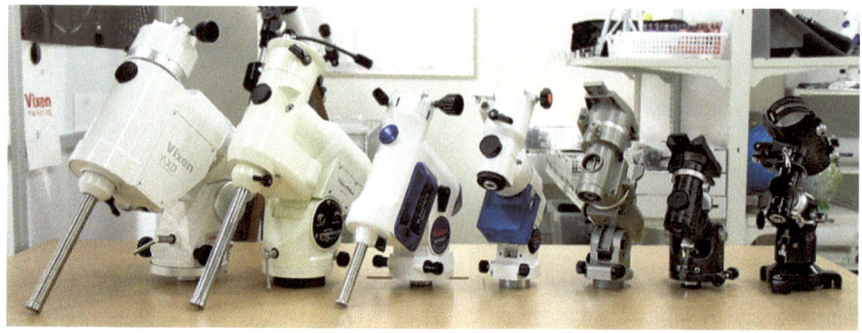

Fig. 2.1 Vixen equatorial mounts family picture (Vixen)

Fig. 2.2 The original Vixen SXW (shown with Vixen ED81S apochromat refractor) (Vixen)

connect to power and the Star Book. The SXW design is elegant, light, clean, and compact. Using the Vixen dovetail system, the Sphinx mount can accommodate a wide variety of optical tubes, either Vixen's or other makes. A variant tabletop version was marketed by Vixen as the SWC, to be used on a special tabletop tripod without counterweight (Fig. 2.3).

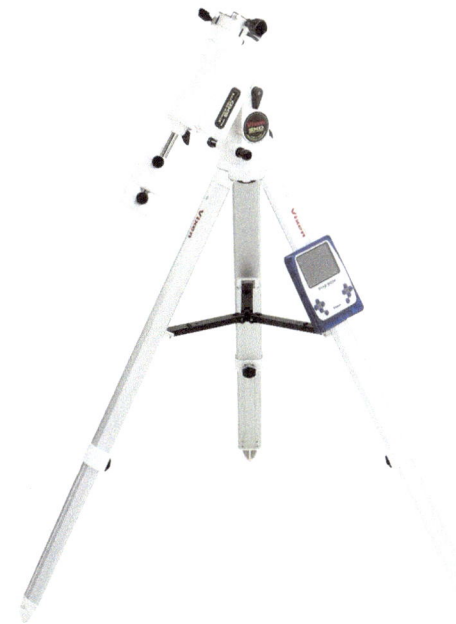

Fig. 2.3 The Vixen Sphinx SXD (Vixen)

The heavier duty Sphinx SXD can be viewed as a SXW on steroids. Using the basic housing as the SXW, the SXD is supplied with the identical Star Book system of electronics and motors as the SXW, but mechanically beefed-up, with the mount, tripod, and counterweights weighing in at 44 lb and a load capacity of 33 lb. The main difference between the SXD and the SXW was the use of improved bearings on the axises, and steel shafts instead of aluminum shafts (Fig. 2.4).

At the top of the line, the original Star Book equipped mounts culminated with the ATLUX model. With the mount, counterweights, and tripod, the ATLUX total weight comes in at almost 75 lb. This increased size allowed the use of optical tubes weighing up to 50 lb to be attached via the Vixen dovetail system. The ATLUX is viewed as a highly stable platform for astrophotography.

Beginning in 2013, Vixen began introducing its updated and upgraded line of mounts. Featuring the next generation Star Book Ten (Ten is not the number 10, but the Japanese word for heavens) computer/controller and new high resolution digital stepping motors, and in ascending order of cost and payload, the Vixen family equatorial mounts currently include the SX2, SXD2, SXP, and the AXD (aka the ATLUX DELUX) (Fig. 2.5).

The SX2 represents the second generation of the entry level Sphinx mount, replacing the SXW in the Vixen line. The quieter digital stepping motors replaced

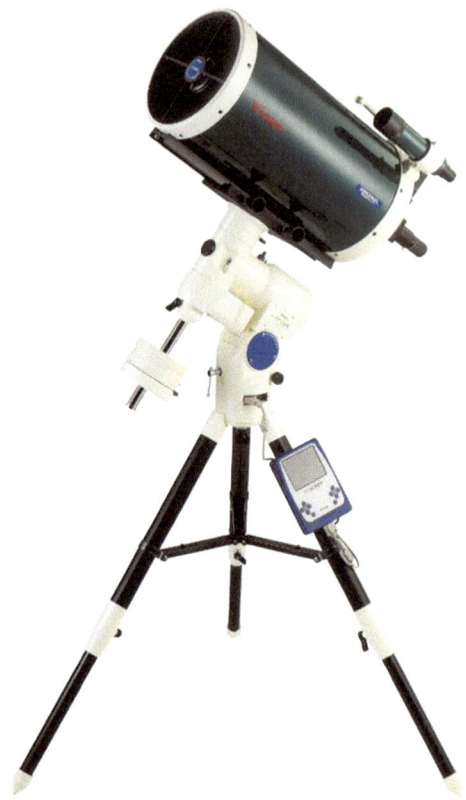

Fig. 2.4 The Vixen ATLUX mount (shown with Vixen VMC 260L catadioptric telescope) (Vixen)

the original DC servo motors, providing smooth, quiet slewing with quick response to commands. The load capacity of 26.4 lb and a mount weight of 27 lb enables the SX2 to be a very portable equatorial mount. The axis shafts are aluminum alloy, and aluminum gears are used for the drive system (Fig. 2.6).

The revised SXD2 Equatorial Mount built on the success of the original Sphinx SXD. The quieter stepping motors replaced the original DC servo motors, providing smooth, quiet slewing with quick response to commands. Heavier components, nine bearings, and increased loading capacity enables the SXD2 to be a solid platform for observing or astrophotography. The SXD2 differs from the SX2 mechanically with use of carbon steel axis shafts and brass wheel gears in place of the SX2 aluminum components. The mechanical upgrades enable the SXD2 to have a load capacity of 33 lb while avoiding any weight gain from the previous SXD model (Figs. 2.7 and 2.8).

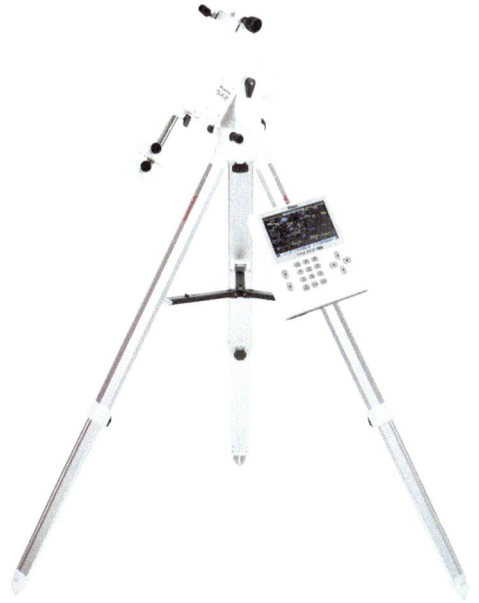

Fig. 2.5 The Vixen SX2 with Star Book Ten (Vixen)

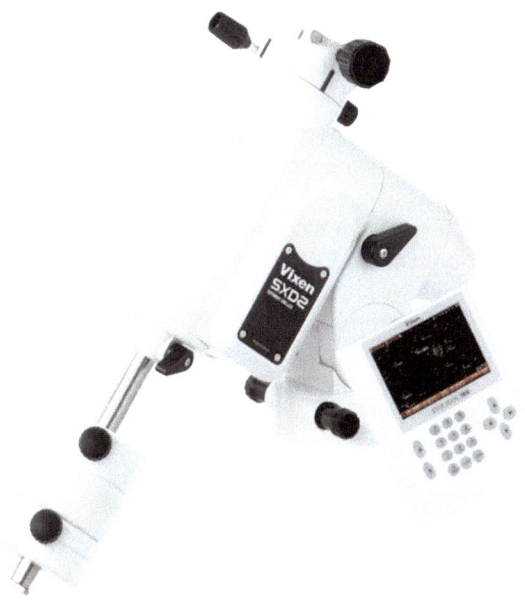

Fig. 2.6 The Vixen SXD2 (Vixen)

Fig. 2.7 Upgraded mechanical parts in the SXD2 (Vixen)

If the SXD2 can be viewed as the SX2 on steroids, the SXP should be viewed as the SXD2 on steroids, a high protein diet, growth hormones, and a lot of gym work. Seriously, the SXP, which stands for the Sphinx Professional, is the ultimate expression of the Sphinx mount lineup and is optimized or astrophotography. Armed with a 40 mm diameter carbon steel declination shaft and low-friction ball bearings, the SXP takes the Sphinx architecture to a load capacity of 35.2 lb, while only being 5 lb heavier than the SXD2 (Fig. 2.9).

The flagship of the Vixen equatorial mounts is currently the AXD, the successor to the ATLUX mount. Sometimes called the ATLUX DELUX, the AXD is the ultimate expression of a Star Book Ten equipped equatorial mount, weighing in at 55.1 lb, excluding counterweights, the pier, or tripod. The AXD is designed for a load capacity of 66 lb!

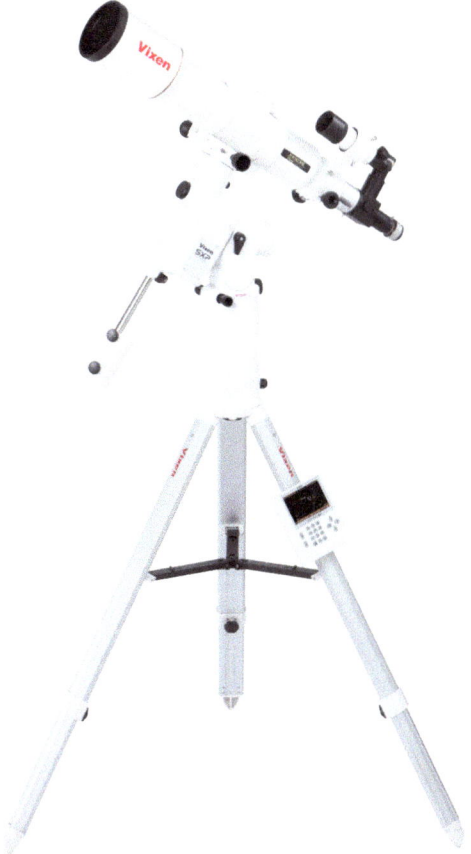

Fig. 2.8 The Vixen SXP (with Vixen AX103S refractor and optional half-pillar) (Vixen)

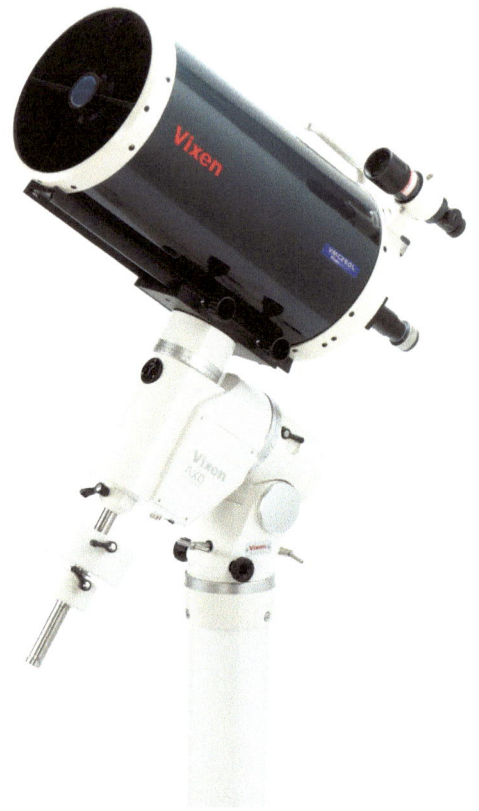

Fig. 2.9 The Vixen AXD (ATLUX DELUX) (shown with Vixen VMC-260L Catadioptric) (Vixen)

Chapter 3

Introduction
to the Star Book TEN

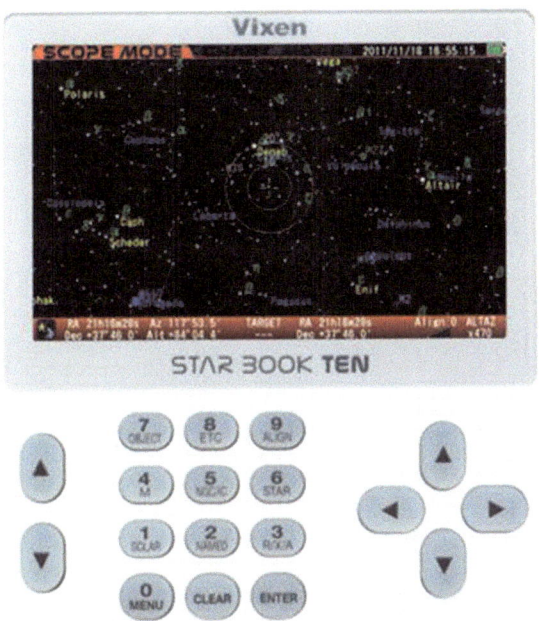

Fig. 3.1 The Star Book Ten (Vixen)

© Springer International Publishing Switzerland 2016
J.L. Chen, A. Chen, *The Vixen Star Book User Guide*, The Patrick Moore
Practical Astronomy Series, DOI 10.1007/978-3-319-21593-8_3

The Star Book TEN represents the latest, most powerful, most capable, and the most current of the Star Book series of Vixen's GoTo computer controllers for the Sphinx line of equatorial mounts. TEN in Japanese means "The Heavens", and is not any reference to the version number of the software or hardware. Designed specifically to drive the new stepper pulse motors that operate at 250 pulses per second (pps), the Star Book TEN provides smooth, quiet, and accurate movement of the Sphinx mount (Fig. 3.1 and Table 3.1).

Table 3.1 The Star Book TEN specifications (Vixen)

CPU	32 bit RISC Processor 324 MHz SH7764
Display	5″ TFT color LCD WVGA (800×480 pixels), with backlighting
Electricity terminal	DC12V EIAJ RC5320A Class 4
Autoguider port	6 pole 6 wired modular jack (for external unit)
LAN port	10Base-T
Mount connector port	D-SUB 9 Pin male plug
Extension slot	For an optional Extension unit (Autoguider)
R.A. and DEC display	R.A.: 1sec. increment, DEC: 0.1 minute increment
Power source	DC12V (Power is supplied from the mount side)
Poser consumption	About 0.25 W (Stand alone unit)
Dimensions	169 mm × 154 mm × 30 mm
Weight	14.11 oz (400 g) excluding the cable and optional extension unit
Celestial object database	272,342 (SAO: 248997, NC Objects: 784; IC Objects: 5386; Messier Objects: 109*; 7 Planets; 1 quasi-planet, the Moon and the Sun) *M40 is a missing number. M91 and M102 are also listed as NGC4548 and NG5866 in the database
Menus and major functions	Automatic GoTo Slewing; Sidereal tracking and different tracking rates for the Sun , Moon, planets, comets, and artificial satellites; Backlash compensation; VPEC; Permanent PEC; Autoguider application; Night Vision Screen; Bilingual interaction; Brightness control, Hibernate control; Built in Speaker; LAN connection updating
RA coordinate display	STAR BOOK Full Color LCD screen; 0.1 min increments
DEC coordinate display	STAR BOOK Full Color LCD screen ; 1.0 arc min sec increments

A frequently asked question is: Can an owner of an older first generation Sphinx mount with an original Star Book upgrade his/her Sphinx mount with a Star Book TEN. The simple answer is no. Since the Star Book TEN is designed to drive pulse motors instead of the original Sphinx DC servo drive motors, the Star Book TEN controller cannot be substituted for the original Star Book used on the original Sphinx mounts. The Star Book TEN, the motherboard within the new Sphinx mounts, bearing design, gear design, and the stepper pulse motors are designed as a system. There is no Vixen update kit available for the older mounts, since the labor and the skills needed for making such an upgrade is most likely beyond the average owner's capability, and if installed improperly can easily compromise the mount's performance. Such an upgrade results in a major rebuild of the mounts, with the labor costs exceeding the cost of buying a new Sphinx mount!

Star Book TEN features a revised control layout from the original Star Book, with a numeric keyboard, with each key doubling as function keys for pull down menus (Fig. 3.2).

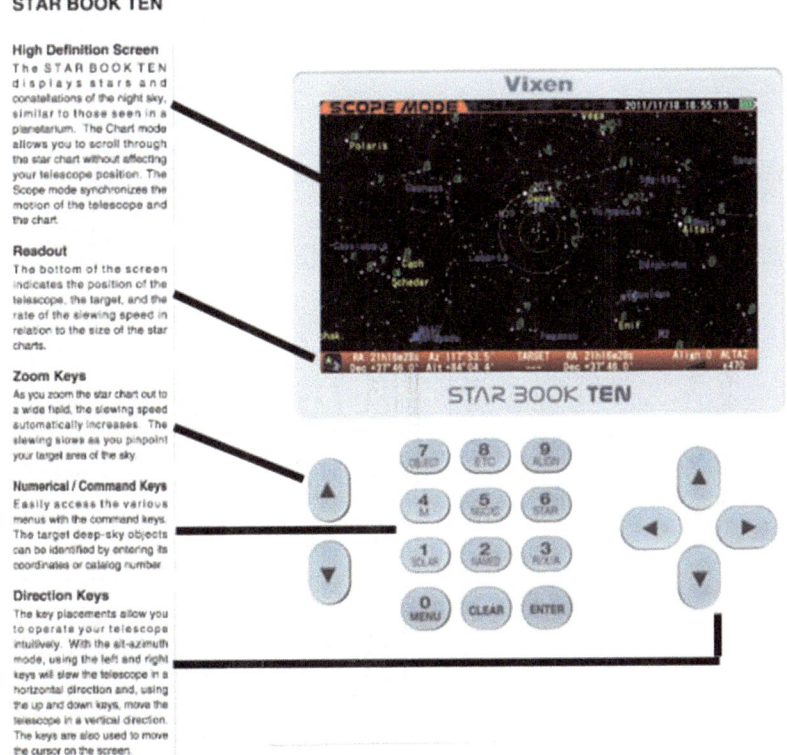

Fig. 3.2 Star Book TEN controller layout (Vixen)

The Star Book TEN plugs into the Sphinx mount through a nine-wire cable with D-SUB9PIN male connectors at both ends. The female connection port for the controller cable is located near the counterbalance bar of the Sphinx mount on the underside of the declination housing. According to the Vixen instruction manual, connect the cable end without the ferrite core at the mount, and the cable end with the ferrite core at the controller. The cable supplied with the author's SX2 mount had ferrite cores at both ends (Fig. 3.3).

Adjacent to the controller cable D connector is the power supply plug. Either a 12 V battery or an AC-to-DC 12-V power supply may be used to power both the

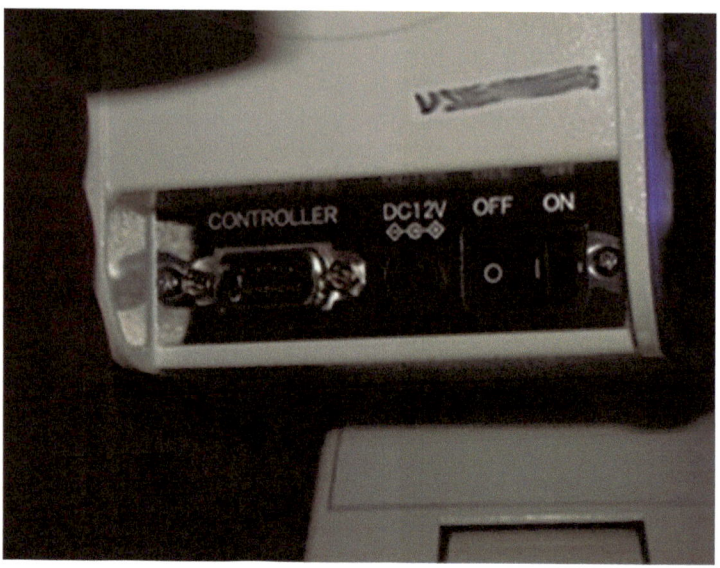

Fig. 3.3 Plug/port layout for Vixen MountsFamily (Vixen)

Sphinx mount and the Star Book TEN controller. The design of the Sphinx on-board motors and motherboard, and the Star Book TEN requires a solid, stable power source. The user/owner of the mount should be aware that low power for a well-used battery supply, or low voltage from the power supply will result in the Star Book TEN not recognizing the mount and its motors. The Star Book TEN will operate only in Chart Mode (Chart Mode will be explained in the next chapter).

The ON-OFF switch is located next to the power supply plug.

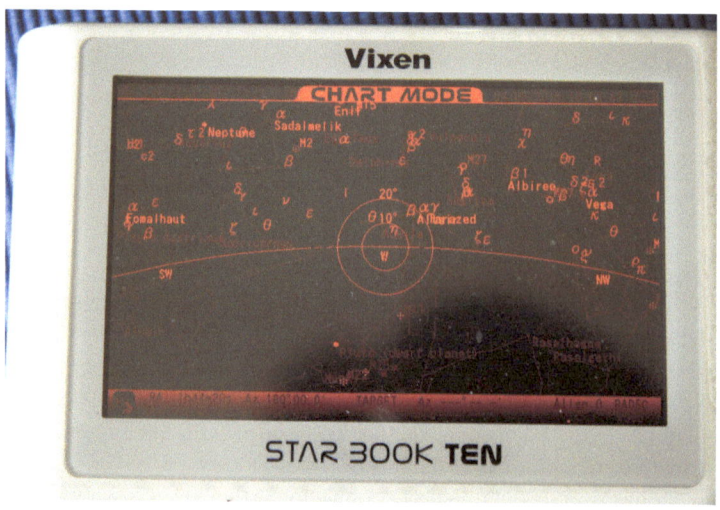

Fig. 3.4 The Star Book TEN in night vision mode (Chen)

As seen in Fig. 3.4, the Star Book TEN has an optional night vision mode, displaying star charts, constellation outlines, and celestial object labeling in night vision saving red illumination (Fig. 3.5).

Fig. 3.5 Example of Star Book TEN in Scope Mode. Note displayed data at the *bottom* of the screen (Chen)

Operating data is displayed at the top of the Star Book TEN display. From right to left:

1. Display Mode: SCOPE MODE: The telescope is linked with the star chart. The telescope follows in the same direction as the star chart is scrolled. CHART MODE: The telescope is independent of the star chart. The star chart is scrolled to select a target object.
2. Date/Time
3. Battery Level (not seen in illustration): Indicated when AC power supply is not used.

Operating data is displayed at the bottom of the Star Book TEN display. From right to left:

1. Tracking ON/OFF icon.
2. Telescope coordinates: Displays the direction of your telescope in Right Ascension and Declination.
3. Altitude Azimuth: Displays the direction of your telescope in azimuth (left and right) and altitude (up and down).
4. Target Name: Indicates a target by number or its common name(within ten characters)

5. Target Coordinates: Displays coordinates of the selected target in Right Ascension (RA) and declination (DEC).
6. Zoom Level Indicator: Levels of zooming up or down the star chart by graph.
7. Motor Speed: Displays a maximum motor speed at a given zooming rate. Up to ×500.
8. Number of Alignment Objects: Increments up to 20 objects used to alignment.
9. Direction Key Mode: Indicates the orientation of the direction keys on the controller keypad as AltAz, RADEC, or X-Y mode.

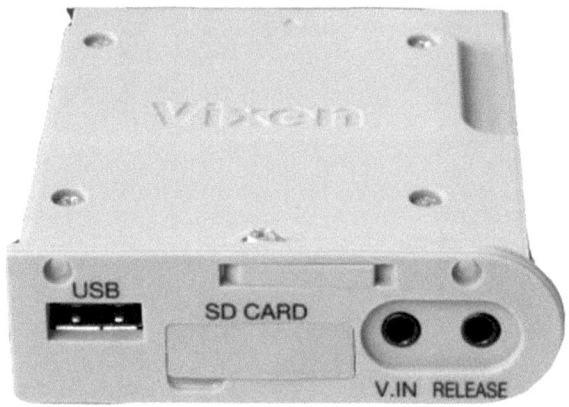

Fig. 3.6 Advance Unit (Vixen)

The Star Book Ten has an optional expansion module called the Advance Unit by Vixen. This unit, shown in Fig. 3.6, attaches to the base of the Star Book TEN and is compatible with the Star Book TEN only. The Advance Unit allows for the use of a CCD autoguider to allow for astrophotography. The Advance Unit also allows for video images from an analog NTSC audio-visual signal to be displayed on the Star Book TEN display screen. Vixen has provided an SD or SDHC memory card slot on the Advance Unit for recording and saving the video images. Additionally, remote shutter release control of a DSLR camera is possible (Fig. 3.6).

Chapter 4

Basic Operation
of the Star
Book TEN

Operating the Star Book TEN successfully is closely tied with the initial setup of both the mount and the initial settings of the Star Book TEN. User mistakes and errors made during the initial setup are commonly the major source of operator problems with all computerized GoTo telescope/mount systems, including the Vixen Star Books.

The Vixen user manual for the Star Book TEN is well written and provides step-by-step instructions for using the Star Book TEN, and the setup of the Sphinx mount. It is recommended that users of the Vixen Sphinx mounts and Star Book TEN read their owners manual. Be familiar with the terminology and the location of controls of the telescope and mount equipment. Assembling the Sphinx mount, attaching the mount to the tripod or pillar, and mounting the telescope onto the Sphinx mount is well covered in the Vixen manual.

On occasion, don't be surprised if the Star Book TEN arrives with an all-Japanese owners manual. This happened to this author! After all, the Sphinx mount and the Star Book TEN are products of Vixen of Japan. When this happens, the owner has a few options:

- Contact the store where the mount was purchased and have them obtain for you an English (or French, or Spanish) version of the manual.
- Contact the importer or distributor to obtain the correct language manual.
- Or go to the Vixen website, and download the pdf of the manual in the appropriate language.

As with any complex device, even with a well-written instruction manual, things can go wrong that takes some time to sort out. That is the role of this book. This chapter will highlight the normal steps for the setup of the Sphinx mount, initializing and operating the Star Book TEN.

© Springer International Publishing Switzerland 2016
J.L. Chen, A. Chen, *The Vixen Star Book User Guide*, The Patrick Moore
Practical Astronomy Series, DOI 10.1007/978-3-319-21593-8_4

Users of the previous generation Star Book will find transitioning to the Star Book TEN easy. The operation of the Star Book TEN is similar to the original Star Book, with the general look and feel familiar but improved. A review of the later chapter on the original Star Book operation reveals the commonality of the two systems. It is best to view the Star Book TEN as an enhanced system with an improved man-machine interface.

Users of other GoTo systems will find the Star Book TEN operation very different from their experiences. There is less reliance by the user on stacks of star charts and volumes of celestial catalogs with the Star Book TEN. The problems associated with searching through a star chart or astronomy book at night with a dim red flashlight are no more. The Star Book TEN encompasses the built-in star chart with electronic control of the telescope and mount system. GoTo searches with the Star Book TEN can be accomplished in two different ways, either using the pull down menus offering selections of planets, stars, and deep sky objects, or just by simply lining up the crosshairs on the Star Book TEN display on a star or object and selecting GoTo.

When in use, there are two ways to hold the Star Book TEN. Users transitioning from the original Star Book to the Star Book TEN will most likely adopt the two-handed approach, holding the controller much in the same way as holding an Xbox or Playstation hand control, manipulating the keys using their thumbs. The new ergonomics of the Star Book TEN also supports holding the device with one hand while operating the keys with the second hand, much like using an iPad or Android tablet computer. Either method works well, and is a up to the user's comfort.

The user is reminded to take time in stepping through the seemingly complex set of procedures in using the Sphinx mount with the Star Book TEN controller. With practice, the process becomes easier and eventually virtually automatic. Don't be surprised that the initial setup that once took an hour to perform eventually morphs into a 10 min exercise. As the old saying goes, "Practice makes perfect."

Polar Alignment of the Sphinx Mount

Before initializing the Star Book TEN, learn how to physically setup the Sphinx mount. One of the capabilities of the Star Book TEN system is that it allows GoTo searching and tracking with any setup of the mount, meaning precise polar alignment is not necessary for the mount to function. In fact, one could setup the Sphinx mount 90° from the Earth's equatorial axis and the Star Book TEN will still function. But doing so is not recommended. Poor polar alignment does not affect the GoTo searching, but can cause less than satisfactory tracking, especially if astrophotography is being attempted. At least a rough "eyeball" alignment will allow smooth error-free operation of the Sphinx and the Star Book TEN. The Star Book TEN firmware does need to be told in the setup routine whether the mount has be precisely polar aligned or not, allowing for the mount to properly compensate in tracking.

The earth rotates and the stars above appear to cartwheel through the sky above around the celestial North Pole. By aligning the telescope to the fixed point in the sky which doesn't move enables the equatorial Sphinx mount and the Star Book TEN to track objects using the right ascension drive motor only. The mount compensates for the Earth's movement and allows the telescope to track the observed celestial object.

In the Northern Hemisphere, the North Celestial Pole (NCP) is located near the north star Polaris at the end of the Little Dipper handle of Ursa Minor. Polar aligning a German equatorial mount, like the Sphinx, is simply the process of aiming the polar axis of the mount at the NCP. For general observing, the simple eyeball polar alignment is sufficient.

For more precise GoTo searches and particularly for astro-imaging and astrophotography, alignment with the using the polar alignment scope of the Sphinx mounts (standard equipment for the SXD2, SXP, and AXD, optional for the SX2) to aim the polar axis at Polaris is recommended. From astro-imaging and astrophotography, the polar alignment precision should be within 5 arc minutes of the true North Celestial Pole. The Vixen polar alignment scope is a 6×20 polar axis scope with an illuminated reticle that mounts within the polar axis of the Sphinx mount. It has a built-in Polaris position scale that achieves an quick and easy polar alignment within 3 arc minutes accuracy. To use the polar alignment scope, move the telescope/mount combination physically until Polaris is within the field of the polar alignment scope. Center Polaris in the crosshairs of the polar alignment scope using the fine adjustment knobs located on the base of the Sphinx mount. Two knobs are provided for movement in the X-axis and a single knob in the rear of the base for the Y-axis adjustment.

The polar alignment scope is standard with all Sphinx mounts, with the exception of the SX2. If the SX2 is not equipped with a polar alignment scope, the polar alignment process can still be easily accomplished for general observing. Align the index markers, known as set position guideposts, located on the side the declination axis and right ascension axis of the mount so that the telescope lines up with the polar axis of the mount. Using the finderscope or red dot finder of the telescope, move the telescope/mount combination physically until Polaris is within the field of the finder. Finally, center Polaris in the crosshairs of the finder using the fine adjustment knobs located on the base of the Sphinx mount. Two knobs are provided for movement in the X-axis and a single knob in the rear of the base for the Y-axis adjustment. For the polar alignment precision within 5 arc minutes of the true North Celestial Pole, there are various methods that are useful to gain precision. The exact position of the North Celestial Pole is 0.9° from Polaris in the direction of the end star in the handle of the Big Dipper, known as Alkaid.

Initializing the Star Book TEN

With the Star Book TEN and a proper power supply or fully charged battery connected to the Sphinx mount, turn on the Star Book TEN and the mount by toggling the power switch to the On position on the mount. After a Vixen logo screen appears on the Star Book TEN display, the following screen then appears (Fig. 4.1).

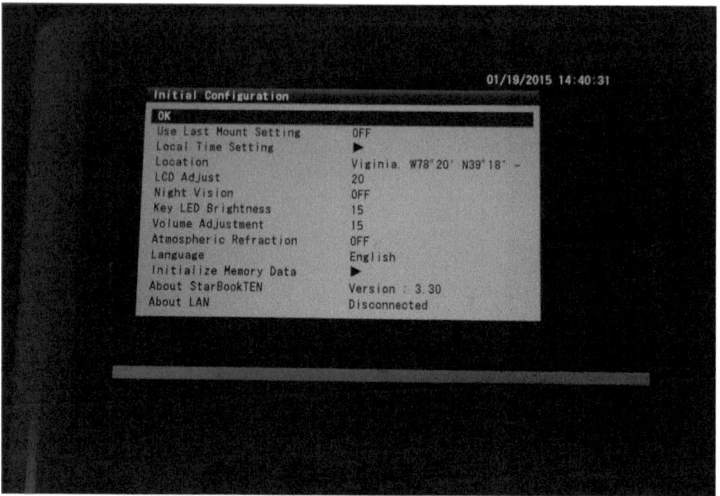

Fig. 4.1 Initial Configuration screen (Chen)

The Initial Configuration screen allows for the setting of the following items in order:

1. If the mount has not been moved, the user can opt to use the previous settings, including the memory of earlier night's objects observed.
2. Local time is set here. Be aware that not only is the time set, but the time differential from Greenwich Mean Time also needs to be set. If the GMT differential is not set properly, GoTo slewing can be very inaccurate.
3. Location coordinates are set here. Note that Vixen requires longitude first and then latitude. If input in the conventional latitude-longitude order, the Star Book TEN and the Sphinx mount will react in an unexpected manner, with the telescope slewing to a ground pointing position for instance.
4. LCD Adjust allows the user to dim the Star Book TEN display to help preserve night vision.
5. Night Vision can be toggled On to change the color display to a red presentation to help preserve night vision.
6. Key LED Brightness allows for the adjustment of the keys.
7. Volume Adjustment allows for adjustment of the sound emanating from the Star Book TEN speaker.
8. Atmospheric Refraction allows for the Star Book TEN to compensate for the refractive properties of Earth's atmosphere during GoTo searches, especially for objects near the horizon.
9. Language settings are accomplished here. The most interesting facet of the Star Book TEN is that it is sometimes delivered in its original Japanese language. The new owner is often surprised when confronted with a Star Book TEN that is unreadable to Western cultures. DON'T PANIC. Fortunately, in its Japanese

state, this line is several Japanese characters followed by a "/" followed by "Lang". Highlight and select this line to access the display language of choice: English, French, German, Spanish, or Japanese. Highlight the language and select. The display instantly converts to the new language.

10. Initialize Memory Data is selected in order to clear the previous nights targets from memory.
11. About Star Book TEN lists the installed firmware version.
12. About LAN informs the user if a successful connection to their LAN is made during firmware upgrades, etc.

Once the initial configuration is addressed and OK is entered, Vixen displays the following warning (Fig. 4.2).

If observing at night, just select Confirm. If attempting daylight observation of the Sun, just select Confirm. By Not Confirm, the Star Book TEN will not proceed forward. This is a Warning screen, probably there to prevent Vixen from legal actions against them when users attempt something stupid, such as pointing the telescope at the Sun without proper safeguards. **Observing the Sun should only be attempted when using solar filters or solar telescopes specifically designed to observe the Sun safely. Attempting to view the Sun without proper precautions and equipment will result in damage to your eyes** (Fig. 4.2).

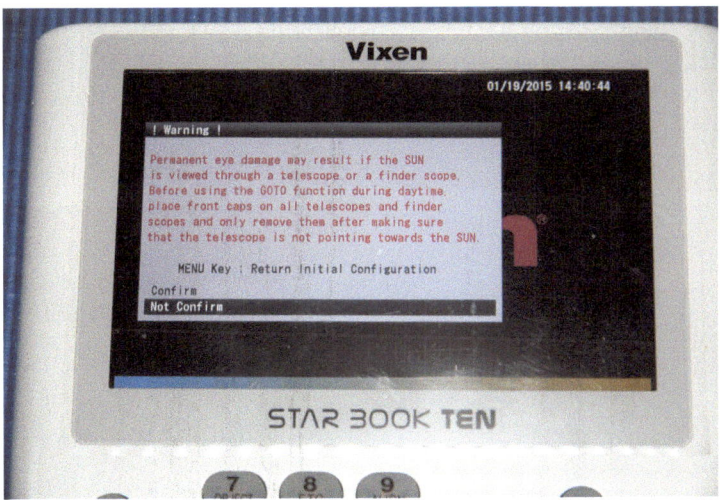

Fig. 4.2 Warning Displaying (Chen)

The next screen is common to both the Star Book TEN and the original Star Book. The majority of new users encounter problems during the setup procedure because of this particular screen. Misinterpreting this screen causes the most problems during initialization procedure for the Star Book TEN and Star Book. The Home position for the beginning of the alignment procedure requires the telescope pointing to the

West. As the example given in the introduction of this book, if the mount and tele-scope is pointed to the East instead of the West, performing GoTo searches during the alignment procedure results in the telescope being aimed towards the ground!

Don't laugh. This happens often. It is easy to see how this mistake occurs. Standing behind a polar aligned telescope facing North, the image on the Star Book TEN display points the user's right. The user mistakenly then points the telescope to their right, unknowingly starting the alignment procedure by pointing to the East in establishing the Home position. A closer examination of the image on the display shows the picture is taken from the front of the telescope and mount with the image being taken in a southerly direction. With the telescope again pointing to the right, but by facing South the right hand direction actually points the telescope to the West.

When standing behind the telescope, facing North, the user needs to know that West is to their left and East is to their right. If standing in front of the telescope and facing South, West is to the right and East is to the left.

Unfortunately, when this mistake is made, the next screen on the Star Book TEN and the original Star Book does not alert the user of this colossal oops. The display will show a sky view map as if the mount has pointed the telescope West. Don't blame Vixen. The Home position identifies to the Star Book TEN where the alignment begins, it's starting point. Up to that point, the Star Book TEN firmware doesn't know where it is pointing. A review of other computerized GoTo systems shows the establishment of a Home position is a common requirement. Some systems require a North pointing Home position, while others share with Vixen the West pointing Home position. Fortunately, once the mistake is experienced and corrected, this mistake never occur again (Fig. 4.3).

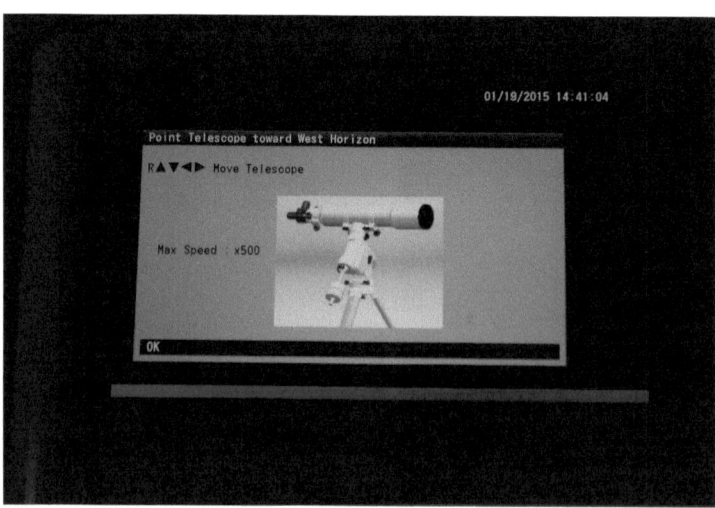

Fig. 4.3 Point Telescope to West Screen (Chen)

With the telescope now pointing in the proper West facing position, the process of selecting stars for alignment purposes begins (Fig. 4.4).

Fig. 4.4 Scope Mode Screen following the Point the Telescope West screen (Chen)

By pressing the number "7 Objects" key, the pull-down Object Menu appears. Most of this menu will be discussed later in this chapter. Of interest in the initializing process is the fifth line on the menu marked "Star". By highlighting and selecting Star accesses the menu of named stars stored in the Star Book TEN memory. Later, during the observing session, the user may access this same menu to select named stars for observation (Fig. 4.5).

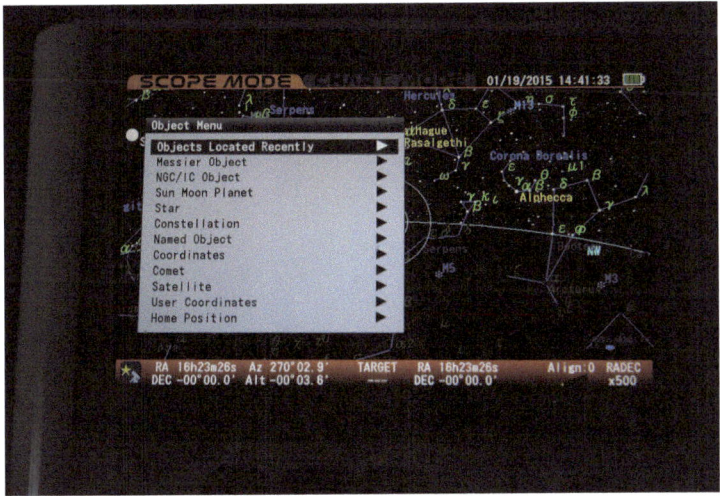

Fig. 4.5 Object Menu (Chen)

The Star Book TEN places a double circle next to the star name for those stars visible at the current time and date. These same stars with the double circle are available for GoTo during the initial alignment (Figs. 4.6 and 4.7).

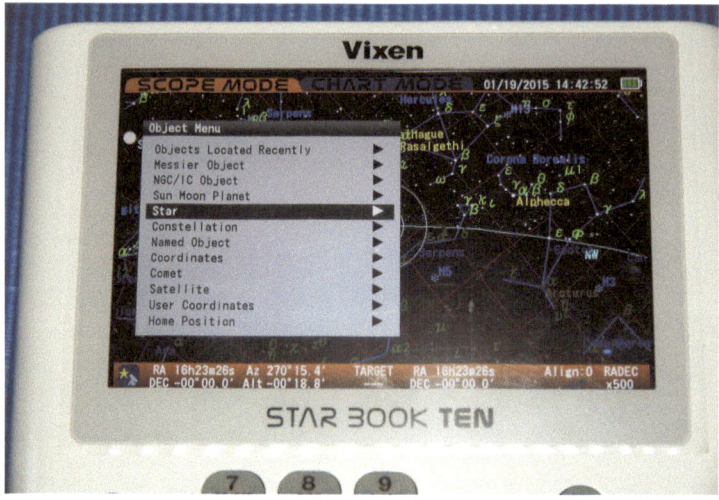

Fig. 4.6 Star Selection from the Objects Menu (Chen)

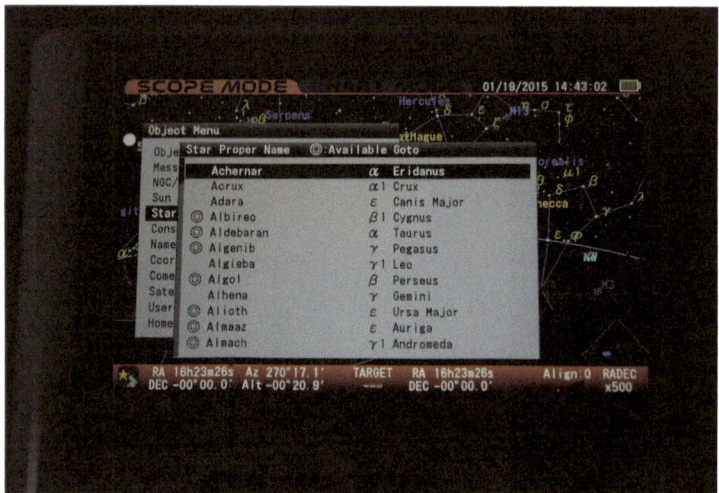

Fig. 4.7 Star Proper Name Menu (Chen)

There is a strategy of selecting stars for alignment. The Star Book TEN firmware allows for up to 20 stars or objects to be used for alignment. The more stars used for alignment, the more accurate the GoTo search. A minimum of two stars are needed for the GoTo searched object to fall into the field of view of a low power wide field eyepiece (approximately 25× to 35× and 65° or greater AFOV). Try to select the first two stars wide apart and to either side of the meridian.

For example, during a winter time observing session, if a bright star in the constellation Orion is selected (such as Betelgeuse or Rigel), the second star selected might be Pollux or Castor in the constellation of Gemini, or Capella in the constellation Auriga.

During the summer months, a good pair of initial alignment stars might be Antares in the constellation Scorpio and Vega in the constellation of Lyra.

As the user, there are a number of good pairings to use from the list of bright stars from the Star Proper Name Menu for the initial two star alignment. The one restriction for selecting alignment stars is that the Star Book TEN firmware requires a separation of at least 10° from each object. It is advisable to chose alignment stars at least 20° above the horizon. The refractive effects of the Earth's atmosphere near the horizon introduces inaccuracies into the Star Book TEN firmware. Also avoid alignment stars at the zenith for a more practical reason. Using straight through finders or red dot finders can cause the user to get into an unnatural and uncomfortable position when viewing straight overhead. A right-angle finder physically solves the positioning, but introduces a disorienting view through the finder. The proper centering of the alignment star can be compromised.

If the observing is to be limited to a particular quadrant of the sky, it is possible to select two stars located in the same region so long as the 10° rule is followed. Returning to the Orion example, Rigel could be selected in Orion, and then Aldebaran in the constellation of Taurus would result in a good alignment pairing. In this example, objects such as the Crab Nebula Messier 1 and the Orion Nebula Messier 42 should easily be found with the eyepiece field of view, although not necessarily centered in the eyepiece. However, a search in the opposite side of the sky will result in a miss. In this example, an quadrant alignment near Orion will have a low probability of finding an object in Ursa Major or Cassiopeia.

As additional stars are added to the list of alignment points, the Star Book TEN becomes more accurate in its GoTo searches. Some users have reported that the threshold appears to be four to six alignment stars or objects are needed to attain center of the field searches. Let experience be your guide.

Vixen goes through great lengths in their instruction manual on the advantages of using pinpoint star images as alignment objects, and not extended objects such as emission nebulas or open clusters. Since there is no definable center of an extended object, this is good advice. The exception are smaller planetary nebulas. Planetary nebulas, such the Ring Nebula Messier 57 or the Ghost of Jupiter NGC 3242, are small enough to be useful as alignment objects. So, by planning the observing session properly, any GoTo searches for small planetary nebulas can also be used for alignment purposes.

A little discussion is needed on the actual process of centering the selected star in the field of view of the eyepiece to establish alignment. Don't be surprised if the mount slewing to the first target star is so inaccurate that the selected star falls outside the field-of-view of the finderscope. It is not uncommon for the Star Book TEN/Sphinx combination to miss the first alignment star by a few degrees, sometimes outside the field-of-view of a finderscope. The challenge becomes using the slow motion controls of the Star Book TEN to bring the target star into the view of the finderscope, and then centering the star in the main telescope eyepiece. Make sure that the star that is sighted in the crosshairs is the named star being used as an alignment star! Things tend to look different through the eyepiece of a finderscope, and users have been fooled into centering onto the wrong star in both the finderscope and the main telescope eyepiece. When this happens, the Star Book TEN doesn't know any better (it's not artificial intelligence), and the alignment is off and all subsequent GoTo searches are in error and will point in the wrong point of the sky. A good method to use to avoid this problem is to select first magnitude or brighter stars for alignment.

Equipping your telescope with a red dot finder or a reflex finder, such as a Telrad or Rigel Quickfinder, will also be of aid. Using a red dot finder and selecting first magnitude stars eases sighting the alignment, especially when the telescope and mount misses the targeted alignment star by several degrees. With an alignment star off-center by several degrees, problems can arise in correcting using the slow motion control direction buttons on the Star Book TEN and zeroing in one the target alignment star. The view in the finderscope eyepiece can be disorienting, depending on the finderscope design (straight-thru or right angle viewing and erect or non-erect image). The red dot finder actually works better for alignment purposes. Remember, the goal of the finder is to get the alignment star into the field-of-view of the main telescope eyepiece, and is not the end-all and be-all of the alignment process. The bright alignment star can still be seen, even when it falls outside of the red dot display window. Use the directional buttons of the Star Book TEN to bring the target star into alignment with the red dot. This process is very intuitive.

Once the alignment star is in the crosshairs of the finderscope or the red dot is on the target star, and assuming the finderscope has been properly aligned, the star should now appear in the field-of-view of the eyepiece on the telescope. Slow the slewing rate down by using the left side Up button, which zooms in the object on the Star Book TEN display while slowing down the slew rate. Center the star in the field of view, and for greater accuracy, change to a higher magnification and center the star again. Many users will use an eyepiece that will give them 100× magnification and is equipped with a reticle to center up the star. Once every thing is centered, hit the align button, also known as the number "9" key. Then go onto the next alignment star.

Occasionally, users have been known to align their GoTo telescopes on the wrong star. The view in the finderscope can sometimes be disorienting. Users of the Star Book TEN, original Star Book, and other competing systems have, on occasion, peered through the finderscope eyepiece and centered the crosshairs on the wrong star! This seems to happen with stars that are dimmer than 1st magnitude. Second and third magnitude stars can present alignment problems, especially when there are similarly bright nearby stars in the neighborhood. For example, by mis-identifying and aligning on Pollux instead of Castor in the constellation Gemini, for example, can cause extreme alignment and GoTo search problems for any GoTo mount. The best advice is don't rush the alignment process and make sure the correct star is sighted in for alignment.

It is recommended that following the setup routine that a few test GoTo searches for obvious and well-known objects be performed to assure the quality of the alignment. During the Winter, obvious objects such as M42 The Orion Nebula or M45 The Pleiades are good choices to make as test targets to make sure everything is working right. The Summer test objects might include easy objects, such as M13 the Great Globular in Hercules or M57 the Ring Nebula. The Autumn test targets might include NGC 869 and NGC 884 the Double Cluster in Perseus or M31 the Andromeda Galaxy. Spring objects might include Mizar in Ursa Major or M51 the Whirlpool Galaxy. The Star Book Ten and Sphinx mount should locate these objects easily when the setup and alignment procedures have been perform accurately. If the GoTo search for these objects result in a complete miss, check your inputs and alignment procedures, then apply the TOTOTA principle: Turn the system Off, Turn the system back On, and Try Again.

Experience has shown that even with care and following procedures to the letter, on rare occasions the alignment is off and erroneous GoTo searches occur. As with any computer program, sometimes for unknown reasons, the computer glitches, and the Star Book TEN is no exception. Apply the TOTOTA principle when this happens. Use a different set of alignment stars the second go around and success should follow.

The alignment process becomes easier with practice. It is common for new users of the Star Book and the Sphinx mounts to initially spent up to an hour on their first setup and alignment procedure. After a few uses of the equipment, the controls and the setup will become familiar and routine, and the setup process becomes a 5 or 10 min process.

GoTo Search Operation

Now that the Sphinx mount is properly setup, the Star Book TEN has been initialized and the alignment process has been completed, the fun begins.

There are basically two methods to perform a GoTo search using the Star Book TEN. The first method is to use the stored pull-down lists of Messier objects, NGC and IC objects, solar system objects, named deep sky objects, and named stars. In general, this is the most popular method amongst owners of Star Book TEN. An alternative method is to use the directional controls and place the crosshairs on the graphical display on an object of interest, zoom-in on the object to provide a more precise centering in the crosshairs, and then activate the GoTo search from the star map. Both search methods work equally well.

It should be noted that adding up the number of objects in the pull-down lists does not come close to the claimed 272,000 objects. The named lists include 109 Messier objects, 7840 NGC objects, 5386 IC objects, and the Sun, Moon, and planets of the solar system. There are 258,977 SAO catalogue stars are available, but not all from any pull-down menu. The Star menu only lists named stars. The remaining stars are precisely plotted on the graphical star map display and can be searched and found by using the crosshair search method.

As a side note: To those who bought naming rights to a star or stars for loved ones as a Christmas gift or birthday gift, these names are not recognized by the International Astronomical Union, or the scientific and astronomical communities. Although a cute and sometimes thoughtful gesture, the money spent on these naming stars is not official and is a pure ripoff of consumers. So don't expect to see your Uncle Floyd or Grandma Sally's name in the Star Book TEN named star database. It won't be there.

By pressing the number 7 key, subtitled Object, the user gains access to the main Object Menu. From here, pull down menus for a number of types of celestial objects can be access for GoTo selections.

At the top of the list allows the user to access a list of previously observed objects. This is a handy method of returning to objects of repeated interest, without having to go through the other menus (Fig. 4.8).

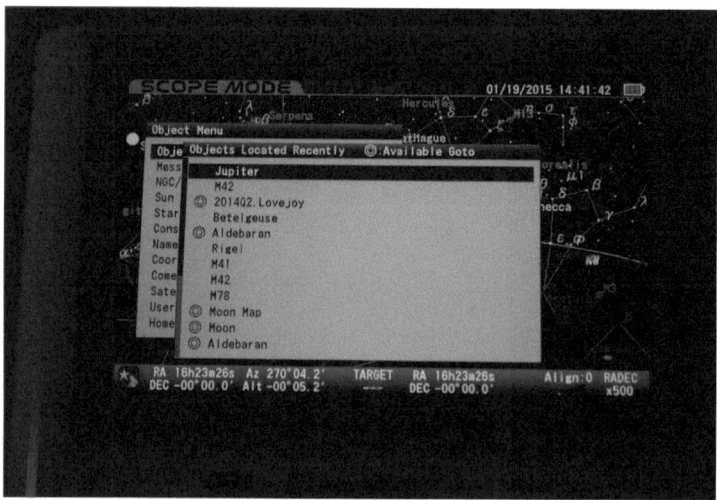

Fig. 4.8 Object Menu—Objects Located Recently (Chen)

By scrolling down the Objects Menu using the down directional arrow on the right side of the Star Book TEN one increment, the Messier Objects are highlighted. Hitting the Enter key selects Messier Objects, and a pull down menu featuring the Messier list is displayed. Alternatively, the Messier list can be accessed by hitting the number 4 key, subtitled M (Fig. 4.9).

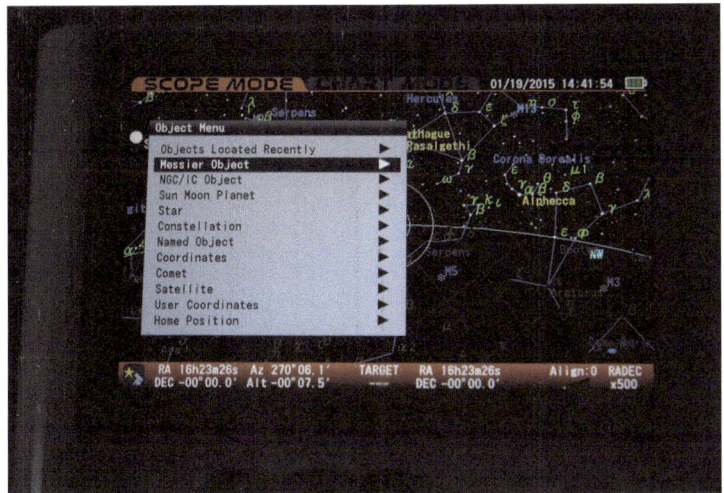

Fig. 4.9 Object Menu—Messier Objects (Chen)

The Messier Object menu presents a listing that contains 109 deep sky objects from the Messier catalog. The Star Book TEN firmware places a double circle symbol next to those objects that are visible at that particular date and time. This is a very convenient method of indicating that certain Messier objects are not visible and are below the horizon.

Scrolling through the list is inconvenient and time-consuming for some users. The alternate and efficient method is once the Messier list is displayed, just type in the object number using the keypad, and the display automatically updates the on-screen list until the full nomenclature has been input. For example, M42 can be accessed by highlighting M1 and then typing in a 4, and then 2, to reveal the highlighted M42.

A select screen will be displayed with pertinent RA/DEC data, magnitude and size data, often accompanied with a picture of the selected object. Highlight "OK" to select (Fig. 4.10).

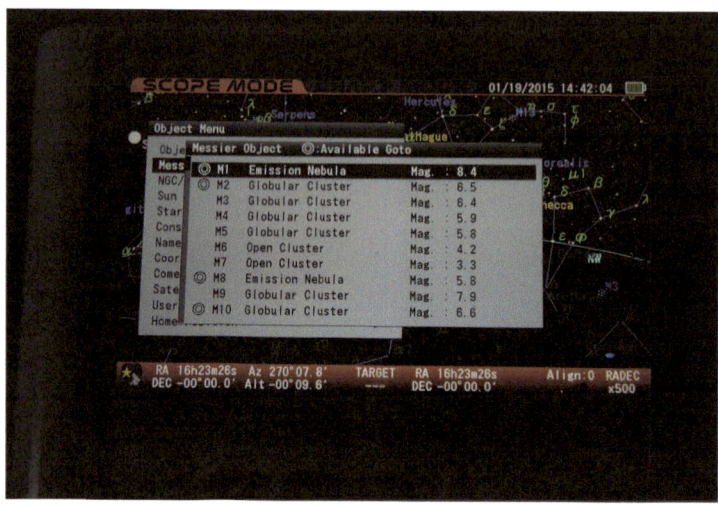

Fig. 4.10 Messier Objects (Chen)

By scrolling down the Objects Menu using the down directional arrow on the right side of the Star Book TEN two increments, the NGC/IC Objects are highlighted. Hitting the Enter key selects NGC/IC Objects, and a pull down menu featuring the combined NGC and IC database is displayed. Alternatively, the NGC/IC list can be accessed by hitting the number 5 key, subtitled NGC/IC (Fig. 4.11).

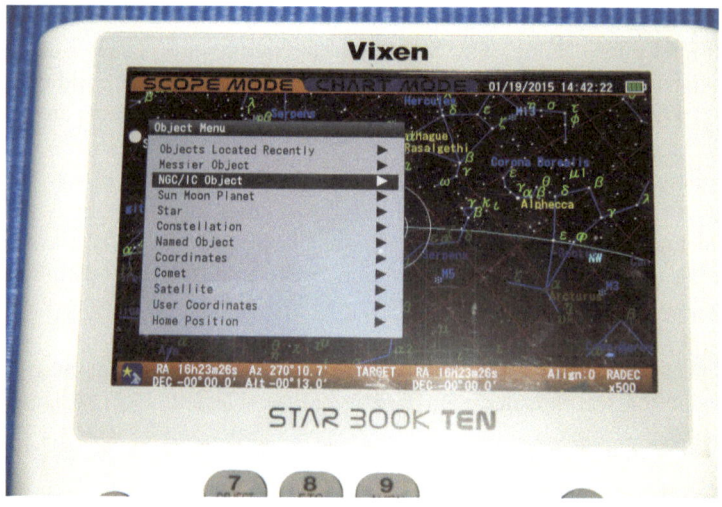

Fig. 4.11 Object Menu—NGC/IC objects (Chen)

The NGC/IC Object menu presents a listing that contains 7840 NGC and 5386 IC deep sky objects from the both catalogs. The Star Book TEN firmware places a double circle symbol next to those objects that are visible at that particular time. This is a very convenient method of indicating that certain NGC/IC objects are not visible and are below the horizon.

By using the menu key on the keypad, the user can toggle between the NGC list and the IC list.

With such a large number of objects in the database, scrolling through the list is inconvenient and time-consuming. The alternate and efficient method is once the NGC/IC list is displayed, just type in the object number using the keypad, and the display automatically updates the on-screen list until the full nomenclature has been input. For example, NGC 2359 can be accessed by highlighting an NGC object and then typing in a 2, 3, 5, and finally a 9 to reveal the highlighted NGC 2359.

A select screen will be displayed with pertinent RA/DEC data, magnitude and size data, often accompanied with a picture of the selected object. Highlight "OK" to select (Fig. 4.12).

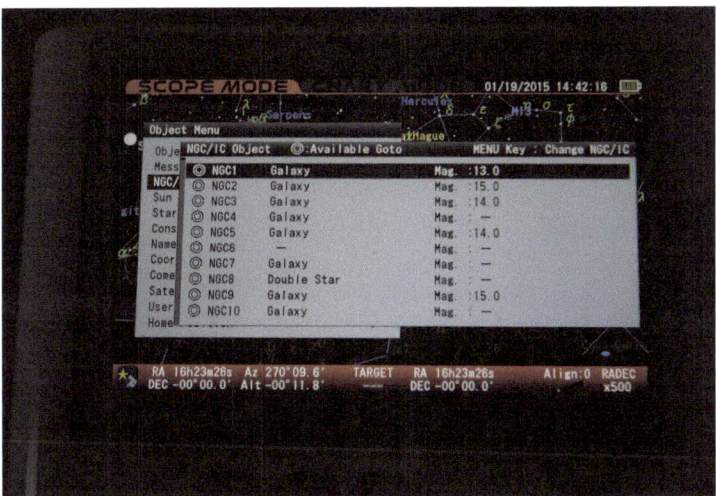

Fig. 4.12 NGC/IC objects (Chen)

By scrolling down the Objects Menu using the down directional arrow on the right side of the Star Book TEN three increments, the Sun Moon Planets are highlighted. Hitting the Enter key selects Sun Moon Planets, and a pull down menu featuring the Sun Moon Planets database is displayed. Alternatively, the Sun Moon Planets list can be accessed by hitting the number 5 key, subtitled Solar (Figs. 4.13 and 4.14a–c).

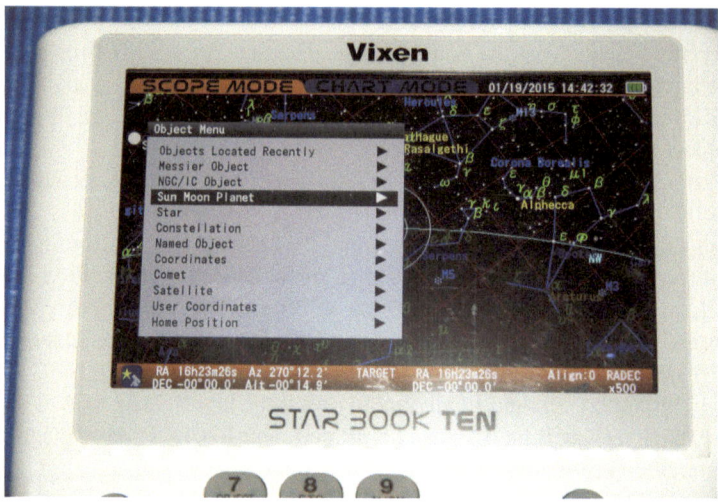

Fig. 4.13 Object Menu—Sun, Moon, Planet (Chen)

The Sun Moon Planets menu lists all the planets including the redefined dwarf planet Pluto. Highlighting the Moon takes the user into a detailed subroutine of the Star Book TEN that includes a visual map of the Moon and listings of major Moon features of craters and Maria (Seas). As the Moon features are highlight, the Star Book TEN will control the Sphinx mounts ever so slightly to move the telescope towards that feature. A very cool firmware feature! (Fig. 4.14b)

When a planet is selected, a photo is displayed on the Star Book TEN screen of the planet with additional specifics of the planet (Fig. 4.14c).

By scrolling down the Objects Menu using the down directional arrow on the right side of the Star Book TEN four increments, the Constellations are highlighted. Hitting the Enter key selects Constellations, and a pull down menu featuring the Constellations database is displayed. There is no direct access to this list from the Star Book TEN keyboard.

The Constellations menu is, at first glance, somewhat odd. A telescope is not needed to sight constellations due to their large angular size. The Constellation function is useful when the Star Book TEN is in Chart Mode and an object within the constellation must be located using the crosshairs search. For example, Epsilon Lyra, the double-double star in the constellation Lyra, is not listed as a named object or a named star on the pull-down menus. Entering the Chart Mode, selecting from the Objects menu Constellations, and selecting Lyra results in the constellation being displayed. Zooming in on the Lyra constellation, using the up key on the left side of the Star Book TEN, allows zooming in on the constellation, with the Bayer designation being displayed for all the stars within the constellation. Placing the crosshair target on Epsilon Lyra and hitting the select key starts the GoTo search (Fig. 4.15).

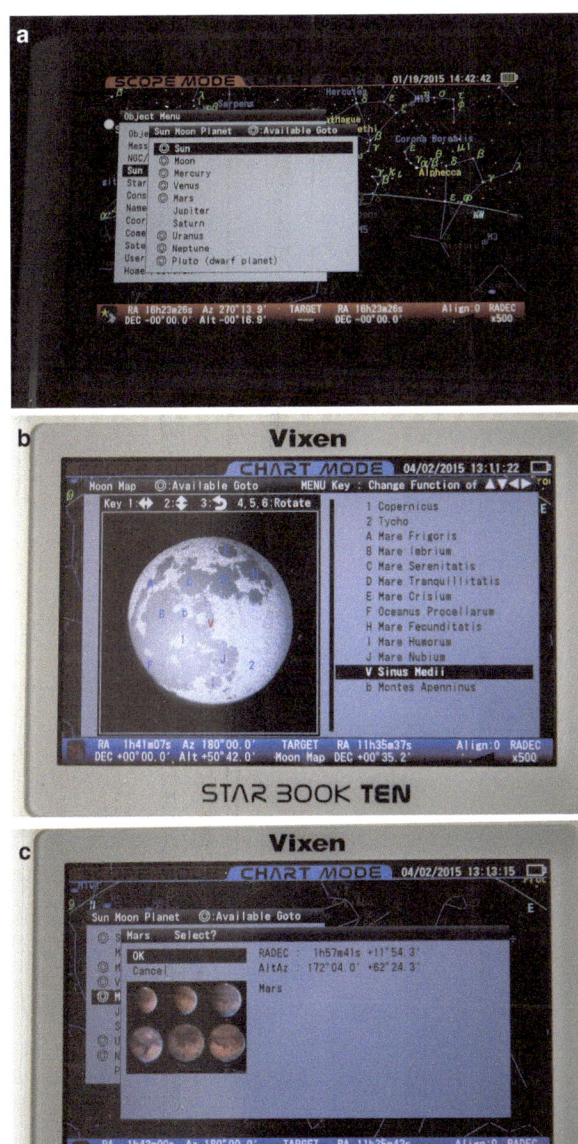

Fig. 4.14 (**a**) Sun, Moon Planet (Chen). (**b**) Moon screen (Chen). (**c**) Example of Planet screen (Chen)

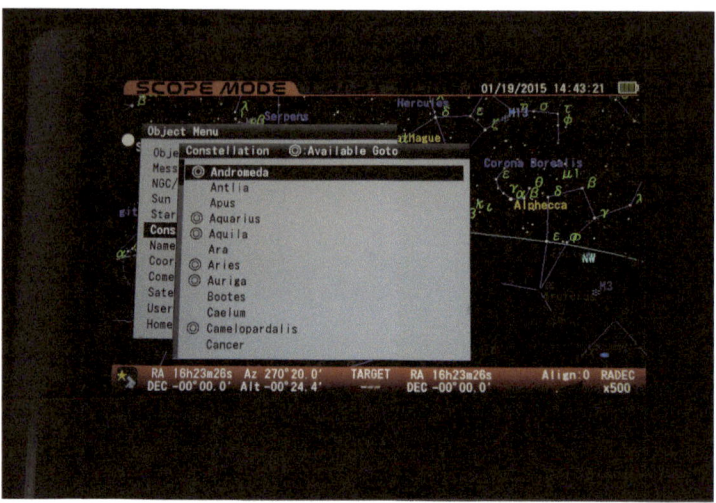

Fig. 4.15 Constellations (Chen)

By scrolling down the Objects Menu using the down directional arrow on the right side of the Star Book TEN to the fifth line item, the Named Objects are high-lighted. Hitting the Enter key selects Named Objects, and a pull down menu featur-ing the Named Objects database is displayed. Alternatively, the Named Objects list can be accessed by hitting the number 2 key, subtitled Named.

Highlighting the object of interest and selecting it is the method of accessing the named object. Again, a select screen will be displayed with pertinent RA/DEC data, magnitude and size data, often accompanied with a picture of the selected object. Highlight "OK" to select.

The Named objects list is limited to the most famous objects. Don't be disap-pointed if an object you are seeking is not listed. Objects such as NGC 2359, Thor's Helmet, NGC 3372, the Carina Nebula, IC 418, the Spirograph Nebula, IC 4406, the Retina Nebula, and the proto-planetary nebula known as the Red Rectangle are not present in the Star Book TEN Named list. To access these objects, some homework by the observer is required, by either knowing the NGC/IC identification number or by the RA/Dec coordinates. The omission of some of these objects may be a reflec-tion of the faintness of the objects and their observability through a telescope likely to be mounted on the Sphinx mount, but this is just a guess (Fig. 4.16).

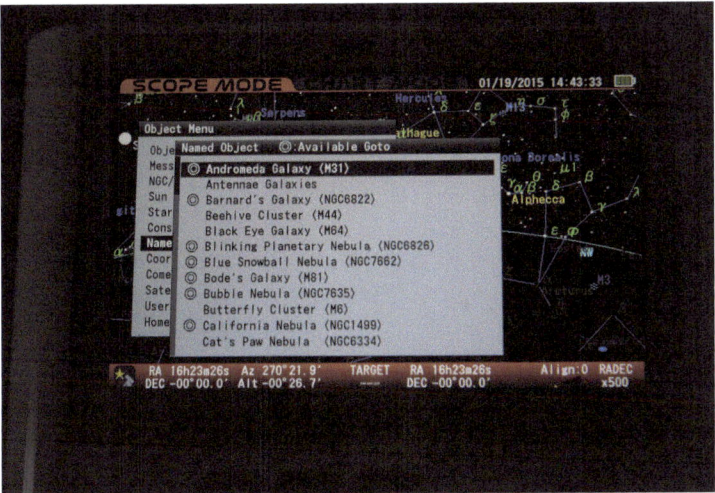

Fig. 4.16 Named Objects (Chen)

By scrolling down the Objects Menu using the down directional arrow on the right side of the Star Book TEN to the sixth line item, the Coordinates is highlighted. Hitting the Enter key selects Coordinates, and a pull down menu showing the right ascension and declination coordinates of an object is displayed (Figs. 4.17 and 4.18).

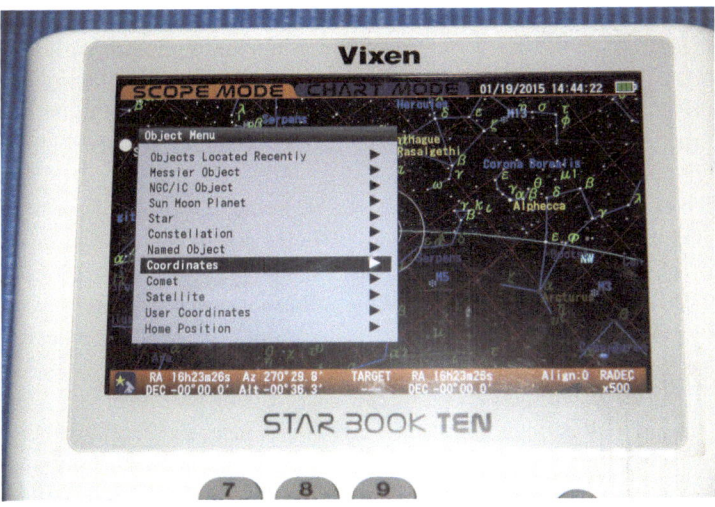

Fig. 4.17 Object Menu—Coordinates (Chen)

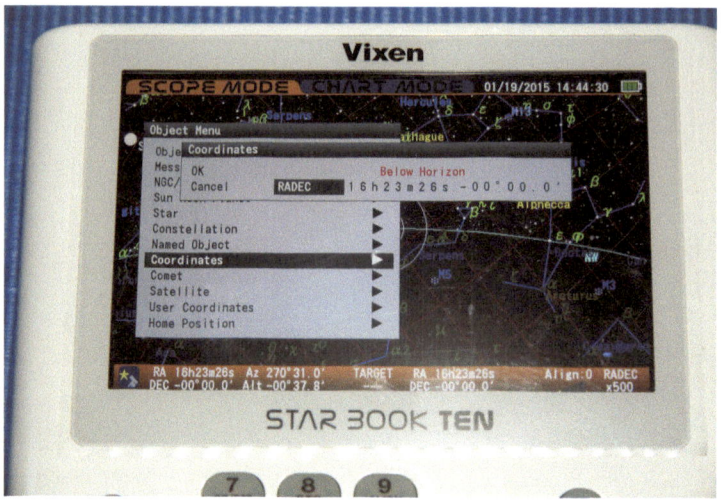

Fig. 4.18 Right Ascension/Declination Coordinates (Chen)

By scrolling down the Objects Menu using the down directional arrow on the right side of the Star Book TEN to the seventh line item, the Comet is highlighted. Hitting the Enter key selects Comet, and a pull down menu enabling the input of orbital elements is displayed. This is an extraordinary capability that has been included in the Star Book TEN, and a significant upgrade from the previous version of the Star Book. From several websites (see Appendix D), the orbital elements of newly discovered or recovered comets can be found for input into the Star Book TEN database and saved. Take time in inputing these orbital elements. The different websites do not follow a pre-described format for presenting the orbital elements data, often times not even labeling the data elements. As a result, some educated guessing is involved in inputing the data into the Star Book TEN, although with some practice this will not be hard. Good pattern recognition will develop in knowing where to input the numbers (Fig. 4.19).

When completed, this capability is one of the most remarkable of the new Star Book TEN. As seen in Figs. 4.20 and 4.21, the illustrated example is of Comet Lovejoy of late 2014 and early 2015. The GoTo search for Comet Lovejoy worked like a charm! The Star Book Ten and the Sphinx mount found and tracked Comet Lovejoy, following the comet's movement against the stationary background of stars (Figs. 4.20 and 4.21).

By scrolling down the Objects Menu using the down directional arrow on the right side of the Star Book TEN to the eighth line item, the Satellite is highlighted. Hitting the Enter key selects Satellite, and a pull down menu enabling the input of orbital elements is displayed. This is another extraordinary capability that has been included in the Star Book TEN, and a capability that was difficult to use in the previous version of the Star Book. From several websites (see Appendix D), the orbital

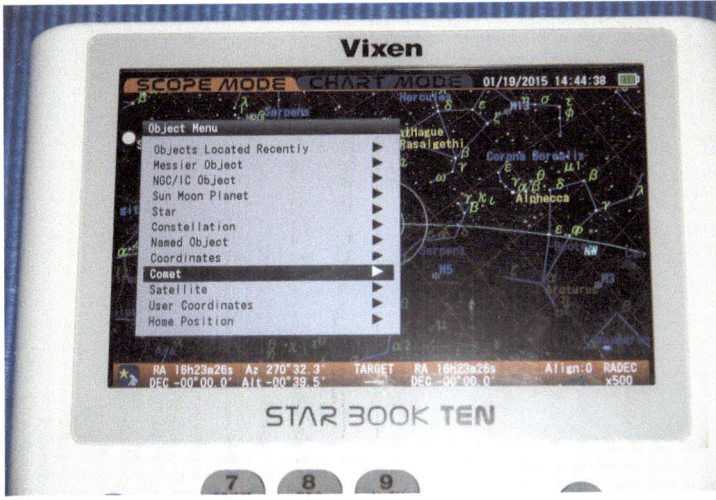

Fig. 4.19 Object Menu—Comet (Chen)

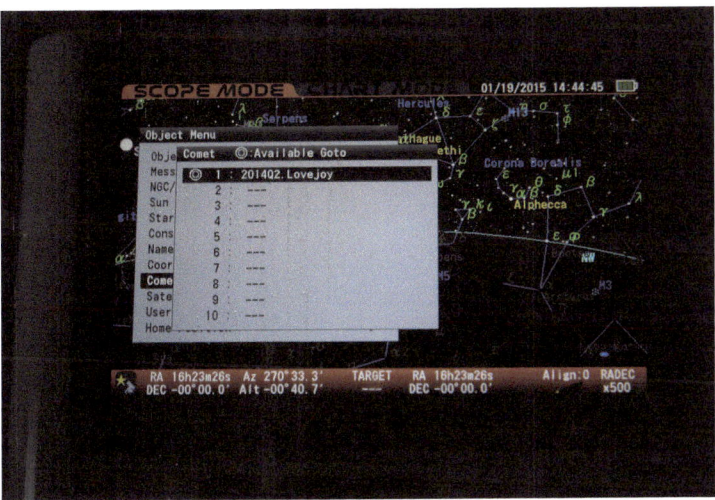

Fig. 4.20 Comet Menu (Chen)

elements of satellites or the ISS can be found for input into the Star Book TEN database and saved. Take time in inputing these orbital elements. Like the previous discussion on inputing the Comet orbital elements, the different websites do not seem to follow a pre-described format for presenting the orbital elements data, often times not even labeling the data elements. As a result, some educated guessing is involved in inputing the data into the Star Book TEN. Again, close study of the website data presentation and good numbers pattern recognition helps (Fig. 4.22).

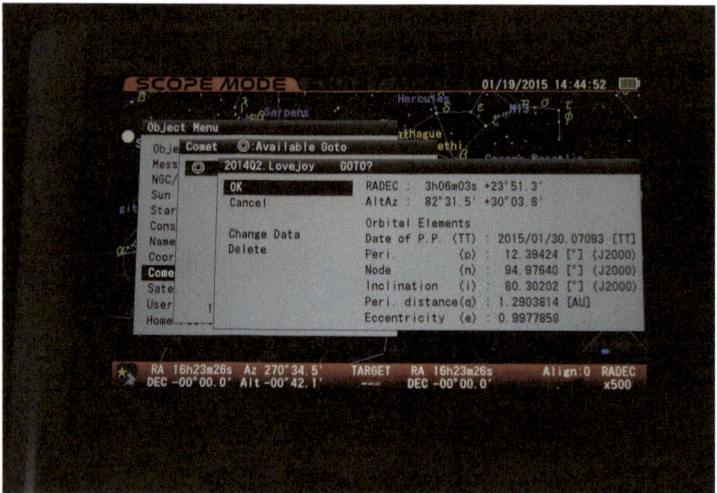

Fig. 4.21 Comet Orbital Elements (Chen)

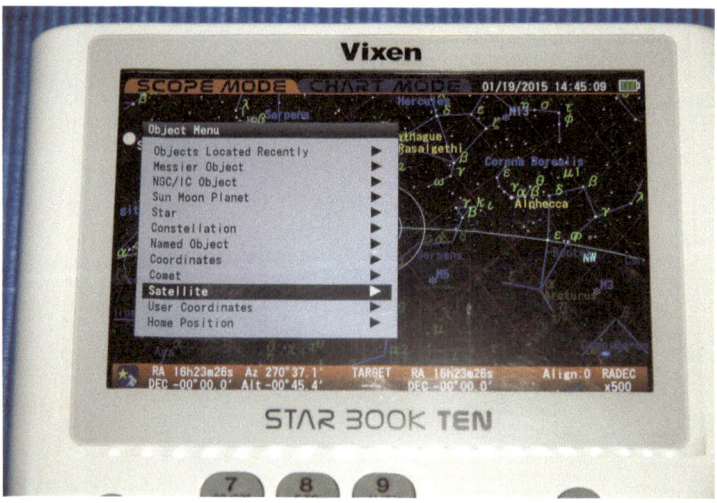

Fig. 4.22 Object Menu—Satellite (Chen)

When completed, this capability is one of the most remarkable of the new Star Book TEN. As seen in Figs. 4.23 and 4.24, the illustrated example is of the International Space Station. The GoTo search for the ISS works like a charm, provided that the user understands when the ISS will be visible across the sky. Some knowledge of the timing of the satellite appearance, as in the example of the ISS, is needed. Otherwise, there is the likelihood the satellite will be shown by the Star Book TEN as being below the dreaded horizon. It should be noted that the ISS orbital elements change periodically, since the space station uses thrusters to boost the ISS to maintain its orbit. If the user is planning to observe the ISS, check on the latest available orbital elements.

Fortunately, in the case of the ISS, there is a website that continuously tracks the ISS, and provides a reasonable timing aid for activating the Satellite GoTo of the Star Book TEN. The website is the ISS Astro Viewer at iss.astroviewer.net (Figs. 4.23 and 4.24).

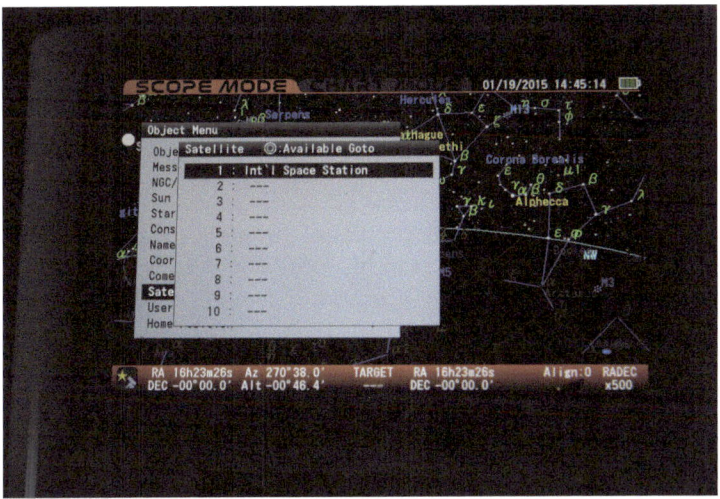

Fig. 4.23 Satellite Menu (Chen)

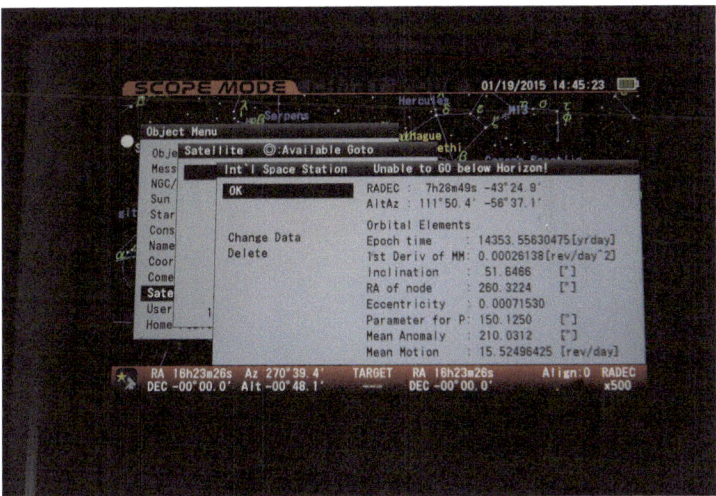

Fig. 4.24 Satellite Orbital Elements (Chen)

By scrolling down the Objects Menu using the down directional arrow on the right side of the Star Book TEN to the penultimate line item, the User Coordinates is highlighted. Hitting the Enter key selects User Coordinates, and a pull down menu selection of previously defined user objects and enabling the input of right ascension and declination coordinates is displayed. This is the user's opportunity to create a custom list of objects not included in the any of the Star Book TEN database.

For example, if Thor's Helmet, NGC 2359, is a favorite target, by highlighting and selecting one of the ten data slots shown in Fig. 4.26, a data input screen appears allowing the user to name the object and input the right ascension and declination information. The now-named object will now appear on this list for future reference. It will not appear on the Named Object list (Figs. 4.25 and 4.26).

By scrolling down the Objects Menu using the down directional arrow on the right side of the Star Book TEN to the ultimate line item, the Home Position is highlighted. Hitting the Enter key selects Home Position and the mount will automatically slew the telescope to the West pointing home position used for the original alignment process. The usefulness of this particular command depends on the user. It is helpful for preparing for the end of an evening's observations. Or if the setup routine needs to be re-initiated due to a power outage, poor initial alignment, or similar event (Fig. 4.27).

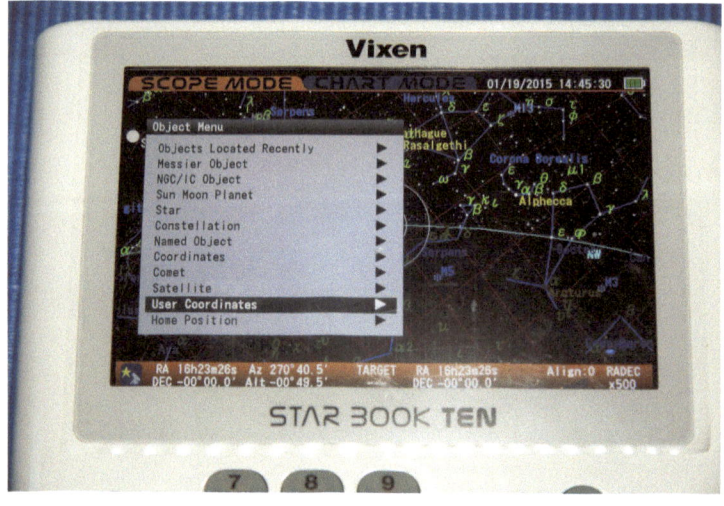

Fig. 4.25 Object Menu—User Coordinates (Chen)

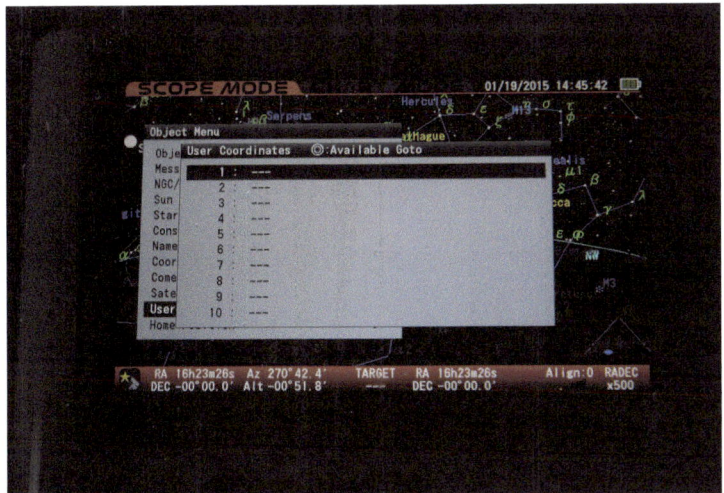

Fig. 4.26 User Coordinates menu (Chen)

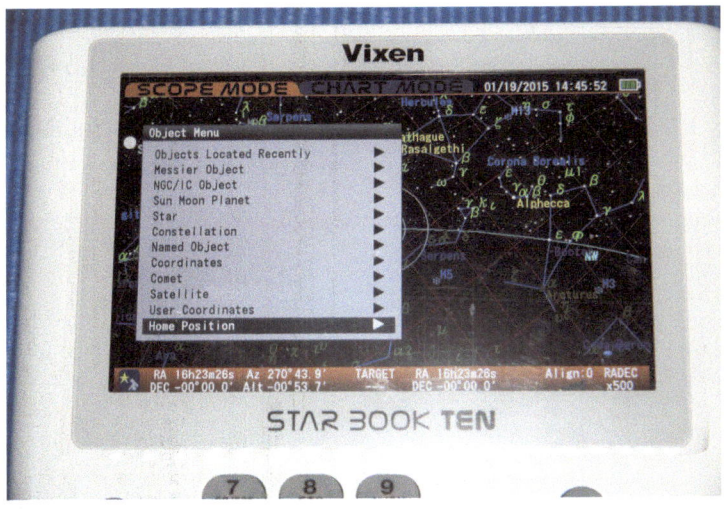

Fig. 4.27 Object Menu—Home Position (Chen)

System Menu

The zero key, "0", when depressed brings up the System Menu, which provides the user with controls for various functions for the Star Book TEN (Fig. 4.28).

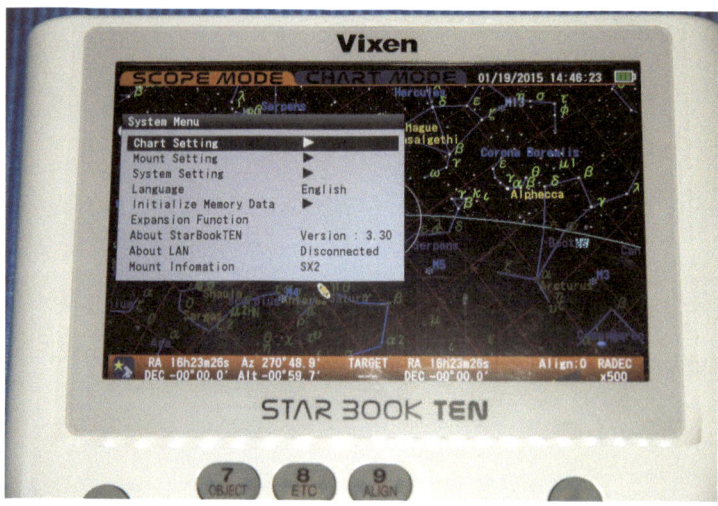

Fig. 4.28 System Menu (Chen)

The System Menu allows the user to customize the operation of the Star Book TEN to suit their personal needs (Fig. 4.29).

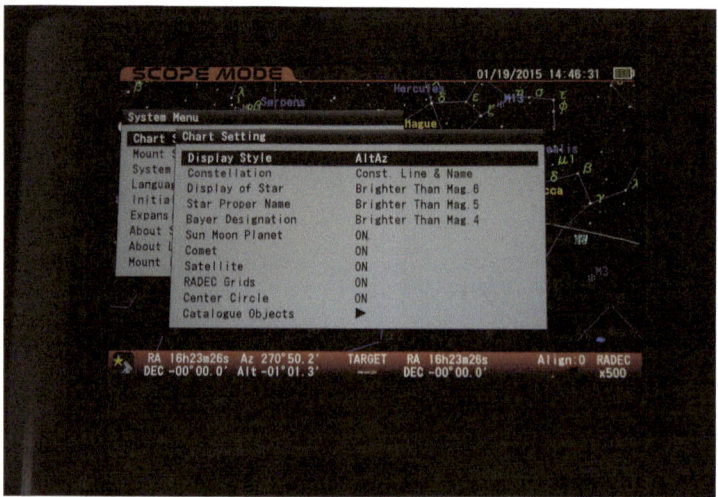

Fig. 4.29 Chart Setting (Chen)

By selecting the first line item on the Systems Menu marked Chart Setting, a number of settings for the display of the star chart are accessed.

- Display Style allows the user to choose between an AltAz or RA/DEC depiction of the sky. The default setting is the AltAz setting.
- The Constellation setting gives the user the choice of having the constellations shown:
  - without lines,
  - with lines,
  - no lines but with the constellation name,
  - lines and name,
  - abbreviated names,
  - lines with abbreviated names,
  - name with abbreviation,
  - and finally lines, name, and abbreviation.
- Display of Star allows the user to have the Star Book TEN to display stars brighter than Magnitude 4, Magnitude 5, Magnitude 6, Magnitude 7, or Magnitude 8. The Star Book TEN defaults to stars brighter than Magnitude 6. The Magnitude 8 setting tends to look very cluttered, but is useful if one is using the crosshair method of GoTo searching for a Magnitude 7 or 8 star. The selection of star magnitudes will not be implemented by the Star Book TEN firmware until the display is returned to the Chart Mode.
- Star Proper name allows the Star Book TEN provide star names for stars of various magnitude settings ranges from Magnitude 1 through 5, always ON, or always OFF. The Star Book TEN defaults to named stars brighter than Magnitude 5.
- The Bayer Designation of stars can also be selected, with similar settings as the previous named stars. The default setting is Magnitude 4 stars or brighter.
- Sun Moon Planet, Comet, Satellite, RADEC Grids and Center Circle are settings that toggle between the labels being ON or OFF. The default setting is ON for all five settings.
- Catalogue Objects allows the user to designate Messier, NGC, and/or IC object labeling, with user magnitude settings for galaxies, globular clusters, open clusters, planetary nebulas, and emission nebulas. The default settings are different for each type of object, and are sufficient for most needs.

By selecting the second line item on the Systems Menu marked Mount Setting, a number of settings for the control of the Sphinx mount are accessed.

- Direction Key allows the selection of how the direction keys on the Star Book TEN operate, between AltAz, RADec, or the mount's own X-Y coordinates. Depending on the user preference, either the AltAz or RADec are the popular choices.
- If an Autoguider is used, the second selection is used to determine the autoguider speed in right ascension and declination. This option is not useable if the Advance Unit has not been installed in the Star Book TEN. The range of speeds are 1–99.
- PEC Record is used to initiate the recording of the periodic error of the drive system. This is only needed when astrophotography is being attempted. By

recording the periodic error, the Star Book TEN drive system can compensate for minute errors caused by the gears of the drive system. All mounts, not matter how well constructed, have minute errors due to manufacturing tolerances. PEC allows for electronic compensation these errors that cause the periodic error.

- Backlash Compensation setting, with a range of 0–99, provides correction for mechanical inaccuracies of the gears and bearings of the Sphinx mount. Again, all mounts, not matter how well constructed, have minute errors due to manufacturing tolerances that cause backlash. The Star Book TEN features an electronic compensation so the backlash is no longer an issue.
- Change GoTo Speed allows for the user setting of the slewing speed, with a speed range from 1 (slowest) to 5 (fastest). The 5 setting saves time during slewing, but also uses power.
- If the mount is equipped with a polar alignment scope, the Polar Scope Light selection allows for the adjustment of the crosshair illumination. The range of adjustment is 0–20, with the default illumination level set at 15.
- Motor power seems like an odd setting at first blush. But there is a purpose for this setting. By selecting from a low setting of 1 to a high setting of 4, the user can control the amount of power used by the drive motors. This enables the user some control over the power usage when using a battery pack.
- Mount Type toggles the control of the mount between an equatorial with polar alignment or no polar alignment. This setting sets the Star Book TEN firmware to recognize and compensate for a greater range of alignment errors. The default setting is without polar alignment. In the default setting, tracking is not just accomplished with the right ascension motor, but also by the declination motor. This mode is sufficient for visual use, but may not be suitable for the intricacies of astrophotography and astro-imaging.
- Cross Over Meridian allows the selection of Warning for Telescope Reverse, Override Meridian stop point and GoTo (East Side). The Vixen Instructional manual details this setting well. The default settings are acceptable and resetting these seems unneeded. For those unfamiliar with the crossover at the meridian, it is an idiosyncrasy of German equatorial mounts that during tracking of an object across the meridian, the mount and telescope needs to be "reversed" or mechanically reset 180° while still pointing to the object. Otherwise, the telescope will strike the mount and cause damage to optics, mount, and drive electronics.
- Delete Align Point Data is when the alignment is off and the user recognizes a flawed alignment star. This function allows for the user to manage of the alignment stars and delete a faulty alignment point. Accessed through the Menu key and under the Mounting Setting line item, a list of Mount Setting controls are displayed. The penultimate line item is "Delete Align Point Data", which when highlighted and selected displays a list of the alignment stars. The user can then delete any faulty alignment star.
- Following Object toggles the tracking motion of the mount. The usefulness of this setting is somewhat vague, with the Vixen Instructional manual describing the alignment of the finderscope as a scenario for its use. Since finderscope alignment can be done with an unpowered mount, toggling on and off the tracking function seems unneeded (Figs. 4.30 and 4.31).

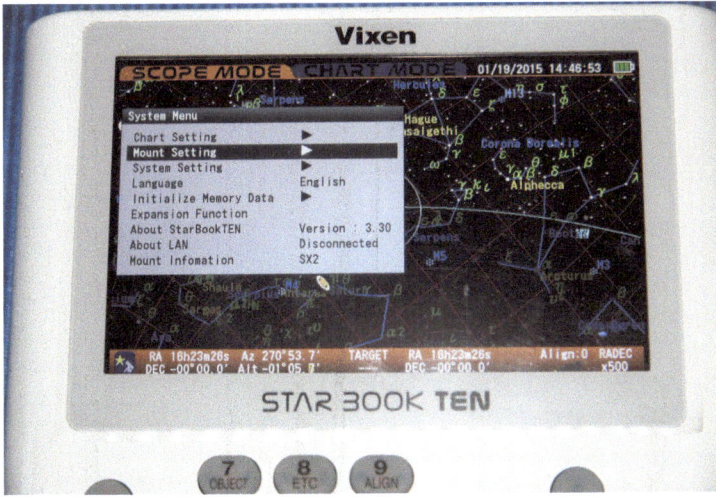

Fig. 4.30 System Menu—Mount Setting (Chen)

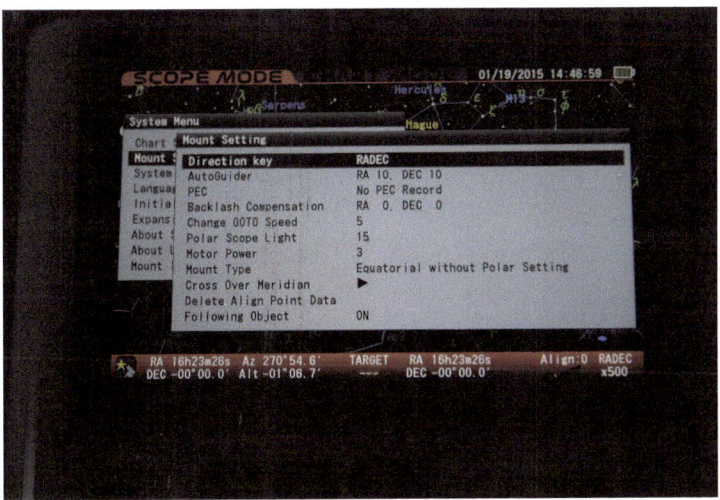

Fig. 4.31 Mount Setting (Chen)

Much of the inputs under System Setting are repeats from the initial settings display when the Star Book TEN is first turned on. Location, LCD setting, Night Vision, Key LED brightness, Atmospheric brightness, Volume adjustment, and GoTo message settings are duplicated here from the opening display. The availability of this settings here are to allow the user to return to these setting in mid-session to make any adjustments necessary due to changing conditions, operations or situations (Figs. 4.32 and 4.33).

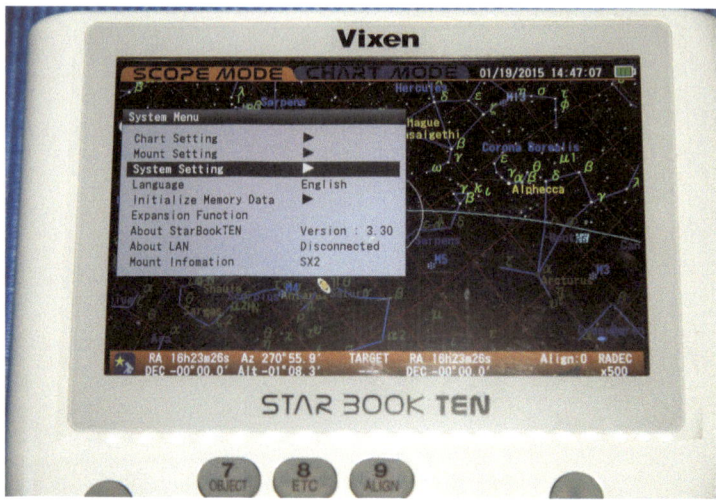

Fig. 4.32 System Menu—System Setting (Chen)

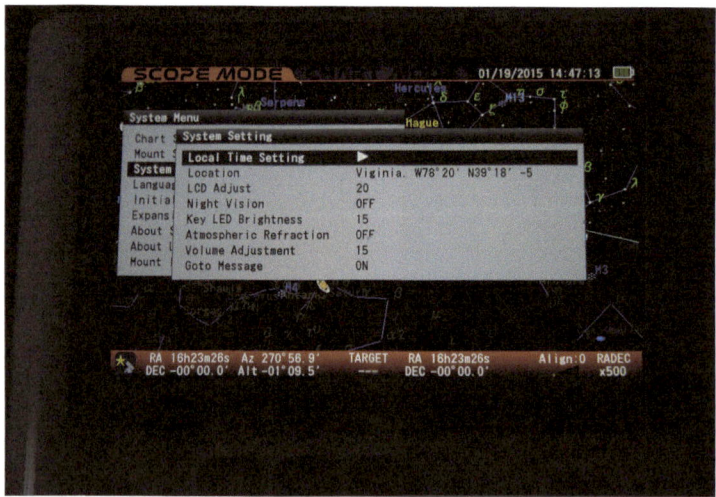

Fig. 4.33 System Setting (Chen)

Again, as in the initialization display, the user can change the language selection of the Star Book TEN in mid-session if necessary (Figs. 4.34 and 4.35).

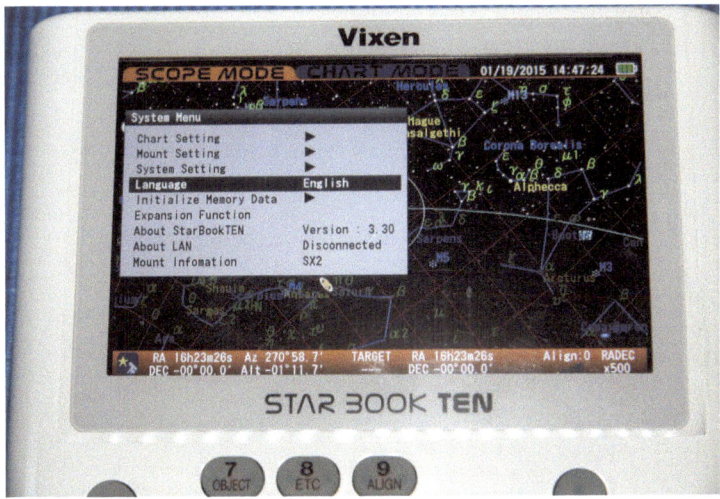

Fig. 4.34 System Menu—Language (Chen)

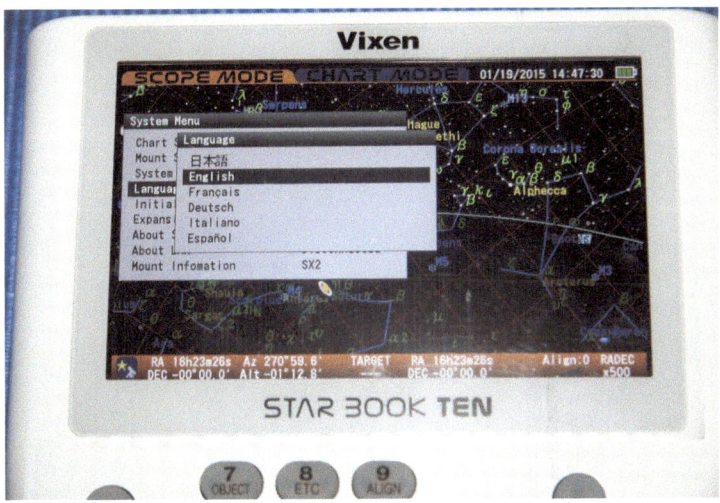

Fig. 4.35 Language (Chen)

This option of Initialize Memory Data is to clear all memory in the Star Book TEN, including the user defined menu items and setup menu items. The only user defined data that is not zeroed out is the local time setting (Fig. 4.36).

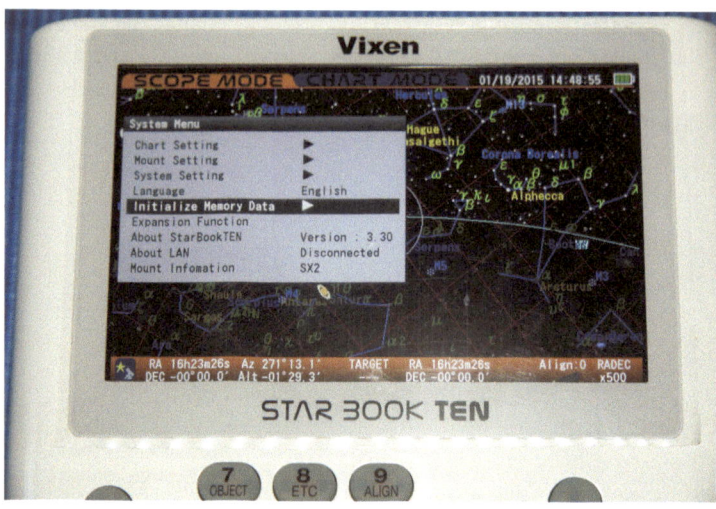

Fig. 4.36 Initialize Memory Data (Chen)

When the optional Advance Unit is installed into the Star Book TEN, this selection activates a new set of setting options that enables autoguider tracking, use of SD card memory, and storage and saving of image data. At the time of the writing of this book, an Advance Unit was unavailable. The on-line manual for the Advance Unit reveals a comprehensive description of how the unit operates. The Advance Unit is useful for those engaged in astrophotography and astro-imaging (Fig. 4.37).

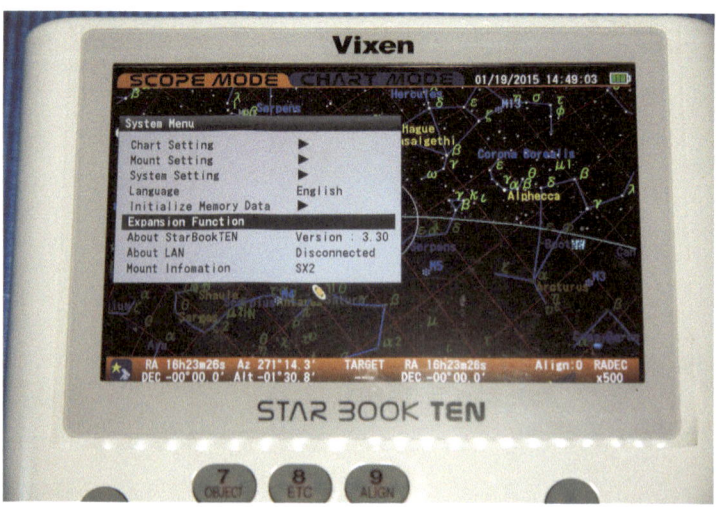

Fig. 4.37 Expansion Function (Chen)

Chapter 5

Learned Lessons in Using the Star Book TEN

Experienced users of computerized GoTo telescopes and mounts are often very adept at jumping and flitting from one telescope to another, one GoTo system to another, without any trouble at all. Just like their high technology cousins the personal computer or the ubiquitous smartphone, there are many commonalities shared among the various systems.

But each GoTo system has its quirks and idiosyncrasies that can prove a challenging as the user climbs the learning curve. The Star Book TEN presents its own set of challenges that with user experience, the operation of the system becomes second nature.

Lessons Learned on Using the Star Book TEN

First and foremost, the Sphinx mounts and the Star Book TEN requires a good source of electrical power feeding into the system. Whether using a Vixen power supply plugged into a wall outlet or a suitable battery pack, the proper voltage and amperage must be supplied to the mount and the Star Book TEN to function properly. The effects of low battery voltage differs between the Star Book TEN and the original Star Book. The first generation Star Book systems manifests the low battery power with a slowing of the DC servo motors and the inability to maintain tracking, with no effect on StarBook controller functionality. The Star Book Ten reacts differently to low voltage power. The Star Book Ten starts up with the setup procedures normally. But after the solar confirmation screen, the Star Book TEN enters and stays in Chart mode. With insufficient power, the mount electronics and the Star Book TEN do not recognize each other, with the normal electronic handshakes and

© Springer International Publishing Switzerland 2016   53
J.L. Chen, A. Chen, *The Vixen Star Book User Guide*, The Patrick Moore
Practical Astronomy Series, DOI 10.1007/978-3-319-21593-8_5

data transfers interrupted. The Star Book TEN automatically enters Chart Mode and will not enter or operate in Scope Mode. The direction keys will not function and the stepping motors do not run, as there is no electronic connection between the Sphinx mount and the Star Book TEN controller. The Star Book TEN behaves as if it is in stand alone planetarium mode, as it is designed to do. Once the user recognizes what has happened, the solution is easy. Either plug in a fully charged battery pack, or connect a Vixen or Vixen compatible wall wart power supply to the mount. Turn on the power switch, and mount and Star Book TEN should then power up correctly. There maybe the necessity to proceed through the setup again, but the mount and Star Book TEN should function correctly.

These are general setup tips that apply to the Star Book TEN, and associated Sphinx mounts:

- Use the most accurate latitude and longitude coordinates for your location. Modern smart phones are equipped with GPS capabilities that can provide the needed lat-long coordinates. If you don't have a smart phone, come on, get with the program. You've got a high-tech telescope mount and a low-tech flip phone? Other sources for accurate lat-long information are car satellite navigation devices, such as Garmin and Tom-Tom, and applications on tablet devices.
- Cell phones and GPS devices are also good sources for accurate time settings.
- The telescope mount tripod should be relatively level. Precision is not necessary.
- Although the Star Book TEN has two alignment options, either equatorial with polar setting or equatorial without polar setting, it is best for accurate GoTo searches to setup as close to True North as possible. Polaris, the North Star, is your friend.
- Choose alignment stars is different parts of the sky, and at least 10° apart. The Star Book TEN allows for up to 20 alignment stars. The more the better, although a minimum of two are needed for initial relatively accurate searches. The Star Book TEN system GoTo searches improve considerably with four alignment stars.
- During GoTo searches, be sure to use a low power wide field eyepiece. A 32–40 mm eyepiece with an apparent field of view of 65° or wider will insure the GoTo search will be successful, with the searched object visible within the field of view.
- The periodic error correction function is only used for astrophotography. For visual use, this function can be ignored. PEC does not seem to play a part in the search accuracy of the Star Book TEN.

Prior to setting up your equipment, make sure of the following:

- Make sure the Star Book TEN and the power source is solidly connected to the Sphinx mount. Many quirky and erratic operations of the Star Book TEN and the Sphinx mount can be traced to loose connections or dirty contacts. A cotton Q-Tip and denatured alcohol can be used to clean grubby contacts with no harm to the equipment and cable connectors.

- Make sure the red dot finder or finder scope is properly aligned prior to entering the alignment procedures. Life becomes easier with good finder alignment.
- During the alignment procedure, lock both axises firmly and never loosen them during your session. All movement of the telescope is done using the electric drive motors of the mount and the Star Book TEN controls.
- Insure that all internal batteries and external power sources are fully charged, fresh, and properly operating.
- Use the manufacturer provided cables. The back rooms of many telescope stores have examples of telescopes and their mounts with malfunctioning electronics and GoTo drives because owners have chosen to use off-brand cables, wiring, and power packs. The false economy of few pennies saved on an off-brand cable will result in tens or hundreds of dollars spent in repairs and hours of lost observing time.
- Protect the Star Book TEN controller from the elements and from dropping on the ground. This is a commercial grade electronic product, and not a military-spec or NASA grade device. There is no conformal coating on the circuit boards to protect from excessive moisture. The Star Book TEN controller does not contain shock mount components. Don't abuse the equipment. Take good care of the Star Book TEN and your equipment will stay away from the telescope backroom awaiting repair.
- If possible, use an inclinometer to set the latitude tilt of the mount. Here again, the handy smartphone will have an application available to use. Place the smartphone on the right ascension axis of the mount and adjust the tilt of the axis to equal the latitude where you will be observing.

Trouble Shooting the Star Book TEN

A number of user problems with the Star Book TEN have their roots in following the setup data inputs and procedures correctly. The most common symptoms of a setup error occurs when a GoTo command causes the scope to point downward at the Earth, or displaying the dreaded "Object is below the horizon" error message when the GoTo object is clearly above the horizon.

The following is a checklist of possible causes of the downward pointing telescope, inaccurate GoTo slewing, or the dreaded error message:

- Home Position screen. Make sure the telescope is pointing West. Then hit OK. If Ok is hit before Home position is achieved, all directional arrows move both axis at the same time. If not pointed to the West, subsequent searches can cause the telescope to point to the ground!
- Substitute a red dot finder or a reflex finder for a finder scope. The initial setup GoTo to major stars tend to miss greatly and fall outside the field-of-view of a typical finder scope. A wider viewing angle is provided by the red dot finder with no inverted image or magnification to cause confusion. 1× magnification will not bring other stars in the view, so the user can zero in on the known bright star, thus yielding a quicker setup time.

- Enter the current location with longitude (east or west) first, and then latitude (north or south). This is backwards for the normal "lat-long" order of coordinates. If the coordinates are input into the Star Book TEN in the wrong order, the Sphinx mount and the Star Book TEN reacts in a sometimes puzzling and sometimes funny manner. GoTo searches will be performed through a good part of the sky, misleading the user that all is functioning well. But parts of the sky will be inaccessible. The telescope could either point to the ground, or the dreaded "Object is below the horizon" error message could be displayed on the Star Book TEN.
- Pay attention to the local time setting and especially to time difference from Greenwich time. Remember a change from Standard time to Daylight savings time results in a 1 h change in the time differential between local time and Greenwich time. Also, the + or − sign for Greenwich Mean Time has a great impact. If the mount is used east of Greenwich, UK, such as Europe or Asia, a "+" sign is used. If the mount is used west of Greenwich, UK, such as North America or South America, a "−" sign is used.
- Use of the Last Mount Setting is applicable only for permanent mounting. Trying to use the Last Mounting Setting can be tricky and is not recommended for moving back and forth from inside the home to the patio or deck, even if the site is marked. The Last Mount Setting is really meant for permanent installations of the mount, and the normal portable or transportable use of the Sphinx mounts can cause the Star Book TEN to act finicky. No axis lock can be loosened when using the Last Mount Setting. There can be no change of equipment that would change the weight, balance, or configuration of the telescope. A good CR2032 battery must be installed in the StarBook Ten controller to maintain the computer memory of longitude-latitude, alignment stars, and past history of GoTo objects. Any variation to any of these parameters can cause the telescope act strangely, such as pointing to the ground during a GoTo search.
- Make sure on the opening setup screen the location (such as the author's home state of Virginia, with proper longitude-latitude) is selected. Otherwise, the StarBook Ten defaults to Tokyo, with the telescope pointing at the ground during a GoTo search.
- Check local time. On rare occasions, the internal clock of the Star Book Ten loses time. Even with a fresh battery. The wrong local time can be the cause of inaccurate GoTo slewing and downward pointing.
- During the initial setup, pick two widely spaced alignment stars in the east, and two widely spaced alignment star in the west. The GoTo search will be relatively accurate at this point, with the searched object falling within the field of view of a low power wide field eyepiece. As the observing evening continues through the evening, the deep sky objects that the observer has found using the GoTo can then be centered in the eyepiece field and used as additional alignment objects to improve the GoTo accuracy.
- Chose alignment stars at least 20° above the horizon. The refractive effects of the Earth's atmosphere near the horizon introduces inaccuracies to the Star Book TEN firmware.

- Avoid alignment stars at the zenith. Using straight through finders or red dot finders cause the user to get into an unnatural and uncomfortable position when viewing straight overhead. Proper centering of the alignment star can be compromised, with the user possibly centering the finder on the wrong star.

- Experience has shown that even with care and following procedures to the letter, sometimes the alignment is off and erroneous GoTo searches occur. As with any computer program, sometimes for unknown reasons, the computer glitches, and the Star Book TEN is no exception. This is when you need to remember TOTOTA—Turn Off, Turn On, Try Again. When glitches occur, switch off the mount and Star Book TEN, turn it back on and try again. Use a different set of alignment stars the second go around and success should follow.

- Be aware the official Vixen specification of the operating temperature for the Star Book is −20 to +40 °C, or −4 to +104 °F. At the lower temperatures, the Star Book will exhibit a non-responsiveness to commands. The lubricant in the Sphinx mount also tends to thicken at extreme low temperatures and wear and stress on components on the Sphinx mount and the Star Book becomes a concern. It is recommended that the Sphinx mount and the Star Book not be used in below single digit Fahrenheit temperatures. Extreme triple digit high Fahrenheit temperatures should also be avoided.

- A major difference between the Star Book TEN and the original Star Book is the newer system allows for the management of the alignment stars. Accessed through the Menu key and under the Mounting Setting line item, a list of Mount Setting controls are displayed. The penultimate line item is "Delete Align Point Data", which when highlighted and selected displays a list of the alignment stars. The user can then delete any faulty alignment point. This capability does not exist with the original Star Book (Fig. 5.1).

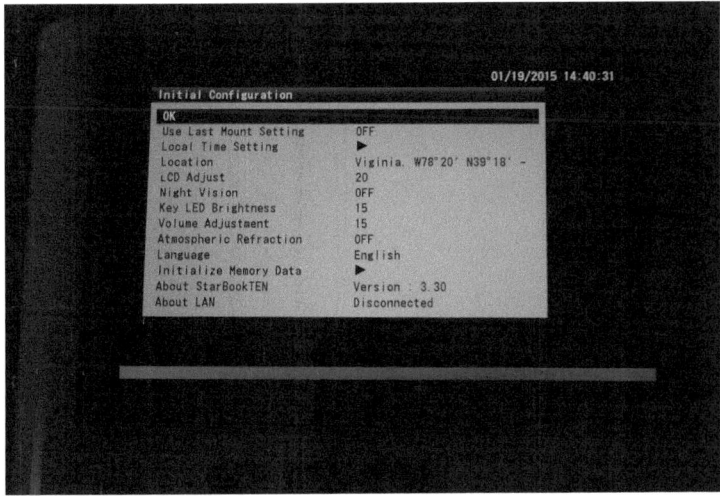

Fig. 5.1 Star Book Ten Initial Configuration screen (Chen)

Many users complained about the original Star Book display being too bright even on the lowest brightness settings. The Star Book TEN offers two solutions to the brightness problem. On the Initial Configuration screen, after the Location data line are two settings that are available to the user to adjust the brightness of the display. The choice of which brightness solution to preserve night vision is left up to the user's preference (Fig. 5.2a, b).

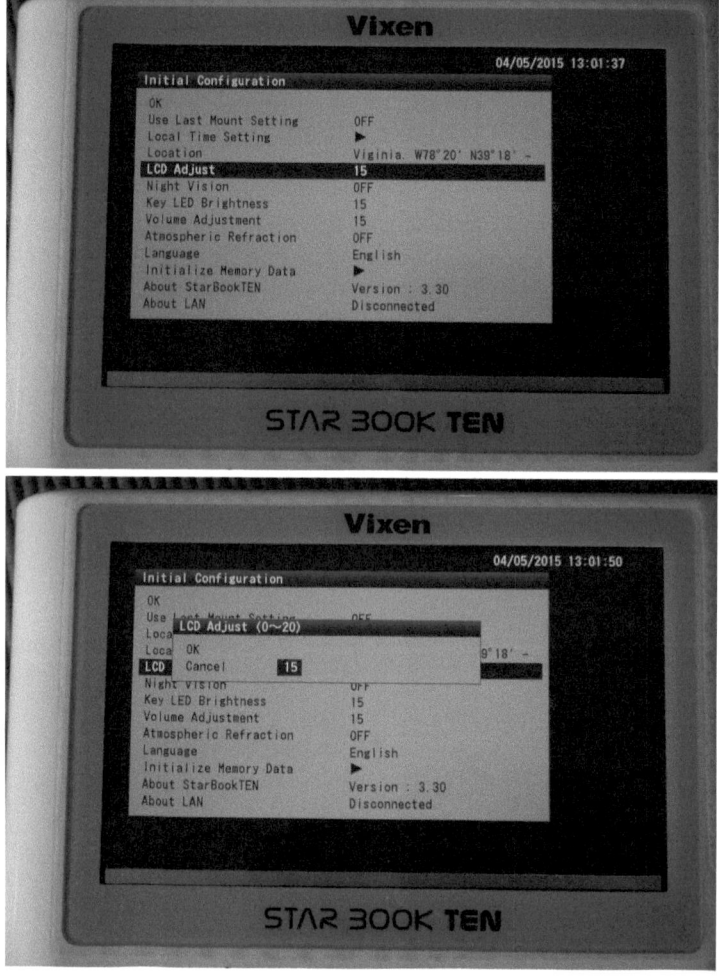

Fig. 5.2 (**a**) Star Book TEN menu screen with LCD Adjust highlighted (Chen). (**b**) Star Book Ten LCD Adjust screen (Chen)

Highlight the LCD Adjust line of the Initial Configuration screen and press Enter. The next pulldown screen will highlight the adjustable level, with a range of 0, which no illumination at all to 20, which is full on illumination. This adjustment allows for use of the full multi-color display (Fig. 5.3a–c).

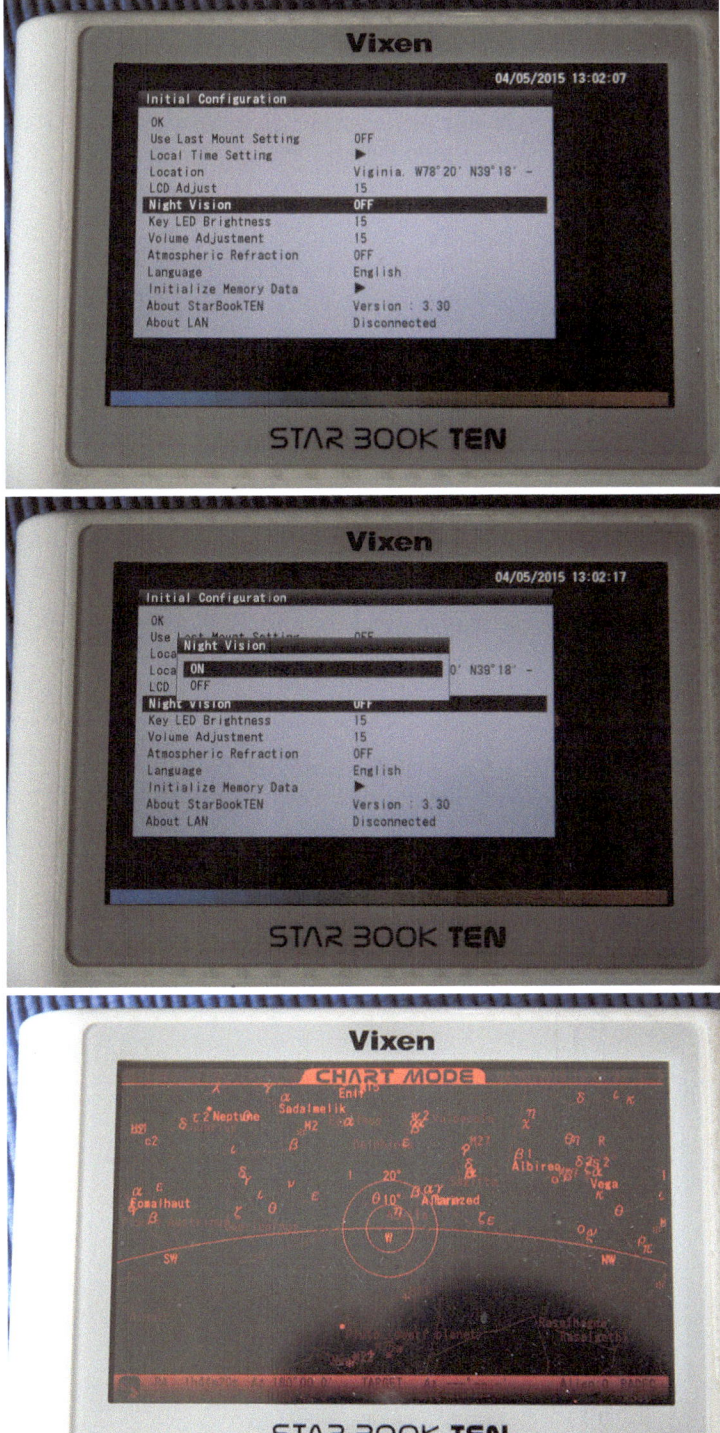

Fig. 5.3 (**a**) Star Book TEN menu screen with Night Vision highlighted (Chen). (**b**) Star Book TEN Night Vision toggle screen (Chen). (**c**) Star Book TEN display with Night Vision on (Chen)

From the Initial configuration screen, highlight and enter the Night Vision menu item yields the ability to toggle on or off the Night Vision display. The resulting amber and black display will save the user night vision, but some loss of chart details are felt by some users.

Chapter 6

Introduction
to the Original
Star Book

Until 2004, Vixen was famous for its German equatorial mounts known as the Super Polaris, and later the Great Polaris mounts. Offered with single axis drive, dual-axis drive, or the remarkable Sky Sensor first-generation GoTo system, the Super Polaris could be found supporting various Celestron/Vixen achromat refractors and Celestron Schmidt-Cassegrain telescopes. Many amateurs bought the Super Polaris to mount many other types of telescopes (Fig. 6.1).

Fig. 6.1 The author's slightly modified Super Polaris mount (note: the custom made counterweight shaft and non-standard counterweight) (Chen)

© Springer International Publishing Switzerland 2016   61
J.L. Chen, A. Chen, *The Vixen Star Book User Guide*, The Patrick Moore
Practical Astronomy Series, DOI 10.1007/978-3-319-21593-8_6

In 2004, Vixen introduced the Sphinx SXW equatorial mount to the world of amateur astronomy. The new mount was a revelation in aesthetic and mechanical design. Until this time, all German equatorial mounts were marketed without drive motors, with motors and electronics sold as externally mounted options. The new Sphinx SXW marked a new approach in German equatorial mounts. It came equipped with internally installed right ascension and declination motors and electronics. A variant table top model was also available, called the SXC, supplied with a diminutive table top base and no counterweights. Otherwise, the SWC is exactly the same as the SWX. The SXW was made available with a lightweight, but heavy duty tripod designated the HAL-130 (Figs. 6.2 and 6.3).

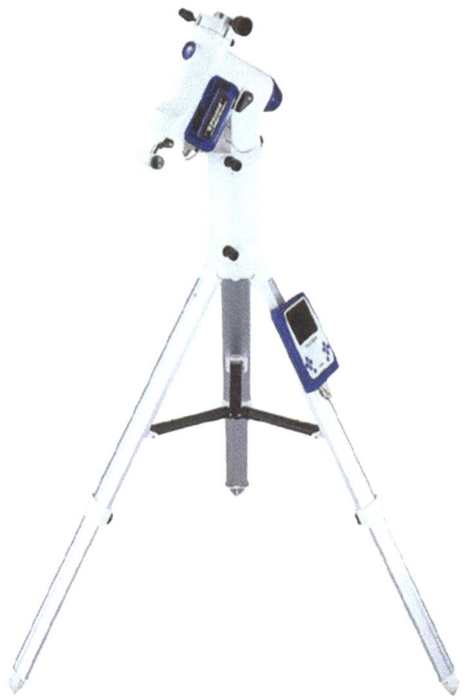

Fig. 6.2 The original Sphinx SXW mount with Star Book, optional half-pillar, and tripod (Vixen)

Along with the new development of the Sphinx SXW came the introduction of the original Star Book.

The Star Book incorporated a 4.7″ diagonal, 320×420 pixel, 4096 color LCD screen and ten control buttons. The screen displayed both a planetarium view of the sky and pull down menus defining the functions of the ten control keys (Fig. 6.4).

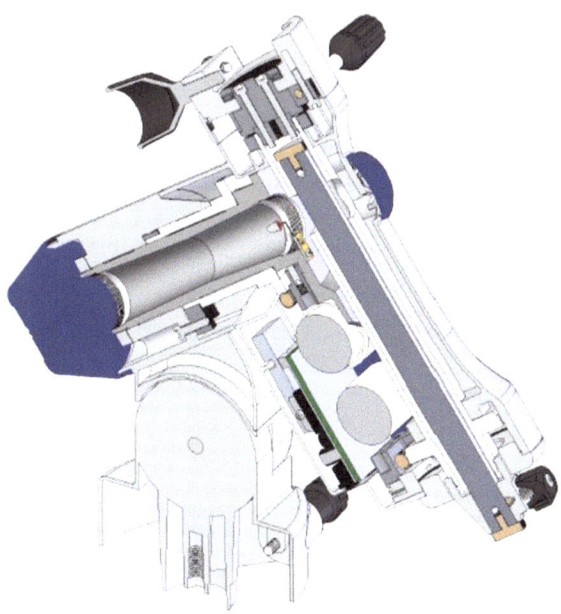

Fig. 6.3 Internal diagram of the Vixen SXW (Vixen)

Fig. 6.4 The original Star Book (Vixen)

Attaching the electrical cables to the Sphinx is easy. Since the right ascension and declination motors and control boards are mounted internally, there are two less cables to carry around, attach to the mount and worry about catching on something when the scope is slewing. A serial cable to the Star Book control unit and the 12 V power cable both plug into the socket panel on the body of the mount just below the counterweight bar. The On/Off switch is also located on this panel (Figs. 6.5 and 6.6).

Fig. 6.5 Connection ports and on/off switch panel on the Sphinx SXW mount (Chen)

Fig. 6.6 Connection ports for the Star Book (Chen)

The Star Book screen presentation was the first implementation of graphical man-machine interface on amateur GoTo mounts. At every zoom setting, the labeling of sky objects was selectable and accurate. The Star Book was powered by a RISC (reduced instruction set computing) processor, and as a result the response time to slews and zooms is remarkably rapid. The graphical display filled the display screen with the upper portion with the star map presentation and a guide to the control buttons on the lower portion. The bottom of the screen contains a legend designating the functions of the control buttons below the screen. Navigating with the control buttons is intuitive and quickly becomes automatic. On the right side of the screen are listed: the current time and date; the RA and DEC where the telescope is pointing; the RA and DEC of the cursor position; and the number of alignment stars that are being used in the current pointing model. None of this auxiliary information interferes in the slightest with the planetarium view of the sky. In fact, the planetarium program is so good that Vixen has enabled the stand-alone use of the Star Book as an independent planetarium using a small 12 V battery pack.

As seen in Fig. 6.7, the Star Book display provides telescope and object data to the user:

1. Scope data: Displays the direction of your telescope in Right Ascension and Declination and Altitude and Azimuth.
2. Target Data: Displays the target identification and target objects coordinates in Right Ascension and Declination and Altitude and Azimuth.
3. Number of Alignment Objects: Increments up to 20 objects used to alignment.

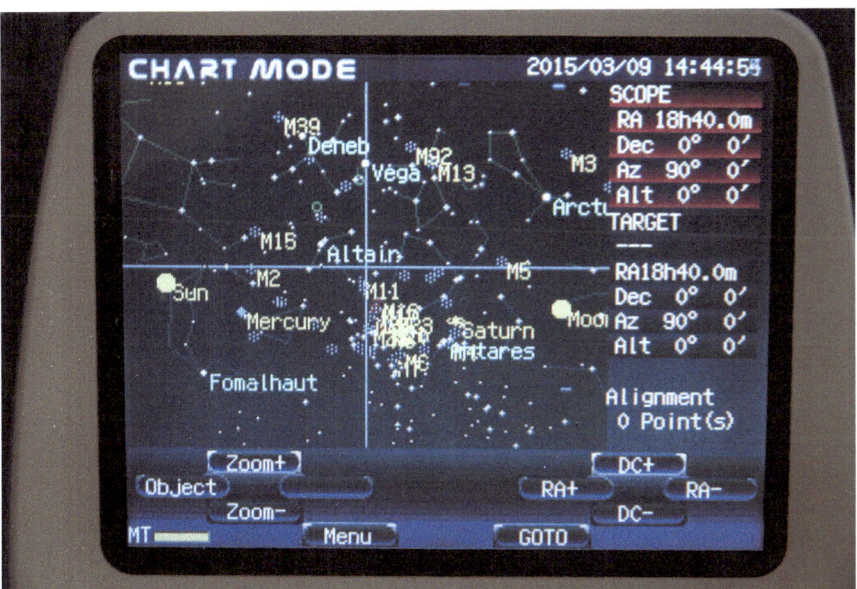

Fig. 6.7 Star Book display example. Note data on *right side* of the display (Chen)

The specifications for the SXW and the first implementation of the Star Book are as follows:

Equipment Type: Computerized German Equatorial Mount
R.A. Slow Motion Axis: 180-tooth wheel gears whole circle movement
DEC Slow Motion Axis: 180-tooth wheel gears whole circle movement
R.A. Coordinates Display: On the screen of STAR BOOK, 0.1 min increments
DEC Coordinates Display: On the screen of STAR BOOK, 1 arc min increments
Polar Axis Scope: Optional
Altitude Adjustment: 0° to 70° (Fine Adjustment with tangent screw: ±15°, 3 step
    elevation)
Azimuth Adjustment: Double screw fine adjustment
Maximum Loading Weight: 16 kg/35 lb
Counterweight: 1.9 kg/4.2 lb
Weight: 6.8 kg/15 lb (without counterweight)

Telescope Controller: STAR BOOK
Processor Type: Cirrus Logic CS89712 32 bit RISC processor 74 MHz
    ARM720TDMI core performance
Display Type: 4.7″ LCD display
Display Resolution: 320×240 pixel resolution with 4096 color
Database: 22,725 objects, Stars: 17,635 brighter than Magnitude 7.0 Messier
    Objects: 110 NGC and IC catalog objects: 4980 brighter than Magnitude 14.0
    Planets, Moon, and Sun
Power Source: DC 12 V

The Vixen component breakdown of the SXW mount, with available options are shown in Fig. 6.8.

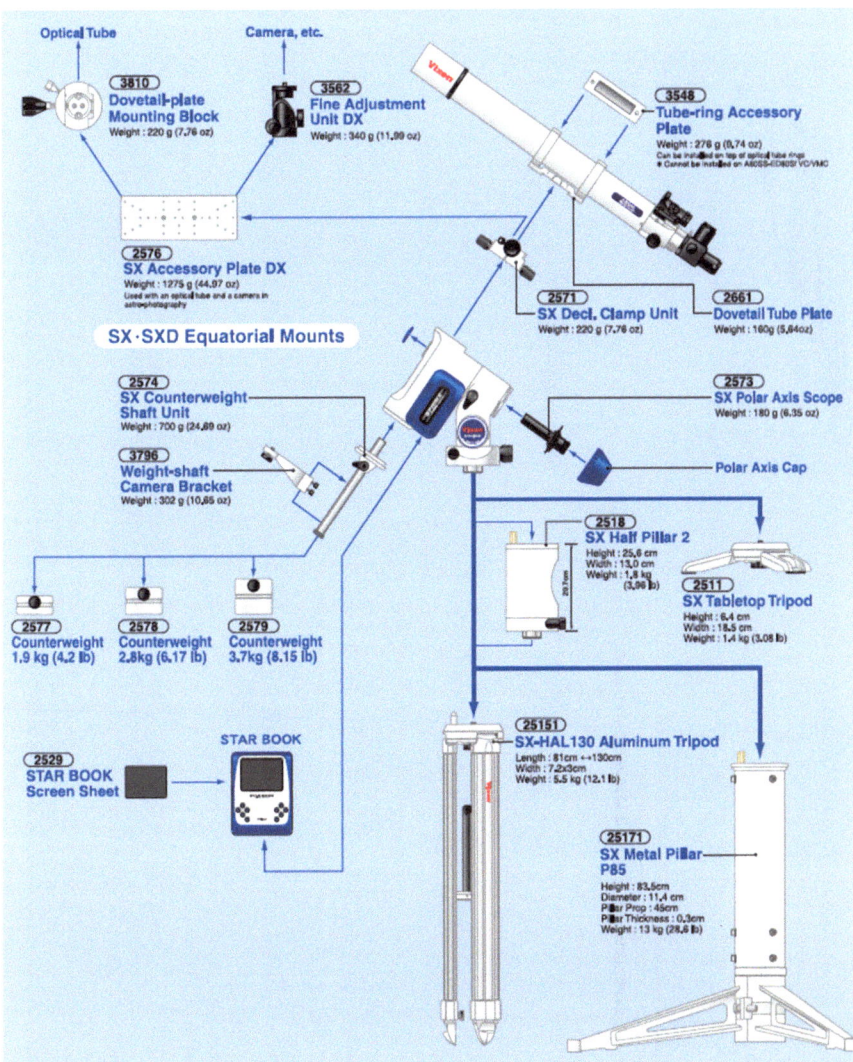

Fig. 6.8 The Vixen Sphinx mount components (Vixen)

Chapter 7

Basic Operation
of the Star Book

The operation of the original Star Book is closely tied with the initial setup of both the mount and the initial settings of the Star Book. Mistakes and errors during the initial setup are the major factors in owners encountering problems with any computerized GoTo telescope/mount system, and the Vixen equipment is no different.

The Vixen user manual for the Star Book is well written and provides step-by-step instructions for using the Star Book, and the setup of the Sphinx mount. It has been reported that although the manual is well prepared, the assembly of the manual is occasionally somewhat flawed. This author has seen examples of the manual where a number of pages were assembled in the manual in incorrect order. The information was all there, just the numbered pages were assembled out-of-sequence. When in doubt, the Vixen manuals can be downloaded from the Vixen website in pdf form. It is recommended that users of the Vixen Sphinx mounts and Star Book be familiar with the owners manual, their telescope, mount equipment, and the Star Book. Assembling the Sphinx mount, attaching the mount to the tripod or pillar, and mounting the telescope onto the Sphinx mount is well covered in the Vixen manual.

A sophisticated device like the Star Book takes time to learn and get comfortable with its operation. Some clarifications are needed to sort out difficulties. Hence, the role of this book. This chapter will highlight the normal steps for the setup of the Sphinx mount, and initializing the Star Book.

There is a commonality between the operation of the original Star Book, and the newer generation Star Book TEN. The users of the Star Book TEN often own the original Star Book mounts, with the general feel of the operation of both controllers being familiar.

As with the Star Book TEN, users of other GoTo systems will find the Star Book operation very different from their experiences. Going out to the field with a Sphinx

© Springer International Publishing Switzerland 2016   69
J.L. Chen, A. Chen, *The Vixen Star Book User Guide*, The Patrick Moore
Practical Astronomy Series, DOI 10.1007/978-3-319-21593-8_7

mount is a simpler and less cluttered operation, with less reliance on stacks of star charts and volumes of celestial catalogs. The problems of searching through a star chart or astronomy book at night with a dim red flashlight are no more. GoTo searches can be accomplished in two different ways, either the pull down menus offering selections of planets, stars, and deep sky objects, or just by lining up the crosshairs on the Star Book TEN display on a star or object and selecting GoTo.

Users of the original Star Book generally adopt the two-handed approach of holding and operating the Star Book. By holding the controller much in the same way as holding an Xbox or Playstation game controller, the keys can be manipulated using the thumbs. The original Star Book does not lend itself to the new ergonomics of the Star Book TEN, which supports holding the device with one hand while operating the keys with the second hand like using a tablet computer. Users who are used to playing video games find the operation of the original Star Book easy and intuitive.

The user is reminded to take time in reading the instruction manual and stepping through the seemingly complex set of procedures in using the Sphinx mount/Star Book controller combination. In fact, practice the setup procedures and GoTo searches indoors prior to taking the mount and controller outside for first light. With practice, the process becomes easier and eventually almost automatic. Don't be surprised that the initial setup that once took an hour to perform eventually morphs into a 10 min exercise.

Polar Alignment of the Sphinx Mount

Before initializing the Star Book, it is important to physically setup the Sphinx mount. A precise polar alignment is not necessary for the mount to function. At least a rough "eyeball" alignment will allow smooth error-free operation of the Sphinx and the Star Book.

The earth rotates and the stars above appear to cartwheel through the sky above around the celestial North Pole. By aligning the telescope to the fixed point in the sky which doesn't move enables the Sphinx mount and the Star Book to track objects using the right ascension drive motor only. Thus the mount compensates for the earths movement and allows the telescope to track the observed celestial object.

In the Northern Hemisphere, the North Celestial Pole is located near the north star Polaris at the end of the Little Dipper handle of Ursa Minor. Polar alignment of any German equatorial mount, such as the Sphinx, is simply the process of aiming the polar axis of the mount at the NCP. For general observing, the simple eyeball polar alignment is sufficient. For astrophotography, an accurate polar alignment is essential.

For more precise GoTo searches and particularly for astro-imaging and astrophotography, alignment with the using the polar alignment scope of the Sphinx mounts (standard equipment for the SXD and ATLUX, optional for the SXW) to aim the polar axis at Polaris is recommended. Astro-imaging and astrophotography, the polar alignment precision should be within 5 arc minutes of the true North Celestial Pole. The Vixen polar alignment scope is a 6×20 polar axis scope with an illuminated reticle. It has a built-in Polaris position scale that achieves an quick and easy polar alignment within 3 arc minutes accuracy. Using the polar alignment scope, move the telescope/mount combination physically until Polaris is within the field of the polar alignment scope. Finally, center Polaris in the crosshairs of the polar alignment scope using the fine adjustment knobs located on the base of the Sphinx mount. Two knobs are provided for movement in the X-axis and a single knob in the rear of the base for the Y-axis adjustment.

If the SXW is not equipped with a polar alignment scope, the process is still easy for general observing. Align the index markers, known as set position guideposts, located on the side the declination axis and right ascension axis so that the telescope is in line with the polar axis of the mount. Using the finderscope or red dot finder of the telescope, move the telescope/mount combination physically until Polaris is within the field of the finder. Finally, center Polaris in the crosshairs of the finder using the fine adjustment knobs located on the base of the Sphinx mount. Two knobs are provided for movement in the X-axis and a single knob in the rear of the base for the Y-axis adjustment. For the polar alignment precision within 5 arc minutes of the true North Celestial Pole, there are various methods that are useful to gain precision. The exact position of the North Celestial Pole is 0.9° from Polaris in the direction of the end star in the handle of the Big Dipper, known as Alkaid.

Initializing the Star Book

With the Star Book and a proper power supply connected to the Sphinx mount, turn on the Star Book and the mount by toggling the On-Off switch to the On position. After a Vixen logo screen appears on the Star Book display, the following screen then appears (Fig. 7.1).

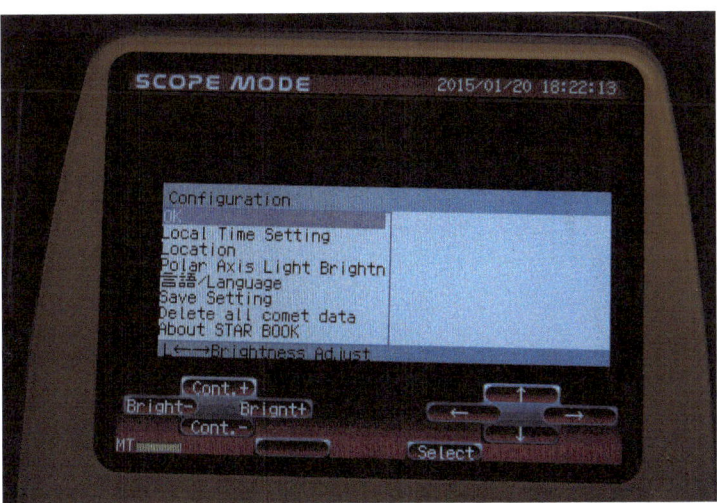

Fig. 7.1 Initial Configuration screen (Chen)

The Initial Configuration screen allows for the setting of the following:

1. Local time can be set. Be aware that not only is the time set, but the time differential from Greenwich Mean Time also needs to be set. If the GMT differential
   is not set properly, GoTo slewing can be very inaccurate.
2. Location coordinates. Note that Vixen requires longitude first and then latitude.
   If input in the conventional latitude-longitude order, the Star Book and the
   Sphinx mount will react in an unexpected manner, with the telescope slewing to
   a ground pointing position for instance.
3. Polar Axis Light Brightness allows the user to adjust the Sphinx polar finder
   illumination to help preserve night vision.
4. Language. The most interesting facet of the Star Book is that it is sometimes delivered in its original Japanese language. The new owner is often surprised when
   confronted with a Star Book that is unreadable to Western cultures. DON'T PANIC.
   Fortunately, in its Japanese state, this line is several Japanese characters followed
   by a "/" followed by "Lang". Highlight and select this line to access the display
   language of choice: English, French, German, Spanish, or Japanese. Highlight the
   language and select. The display instantly converts to the new language.
5. Save Setting allows for the saving in the permanent memory of the time, date,
   and location information. This is valuable for maintaining the location information of your main observing spot. If the mount is moved and used at a temporary
   locale, such as a star party or friend's backyard, don't use this command.
6. Delete all comet data is a self-defining command.
7. About Star Book lists the installed firmware version.
8. Using the left side keys, the left-right directional keys adjust the brightness of the
   display.

Once the initial configuration is addressed and OK is entered, Vixen displays the
following warning (Fig. 7.2).

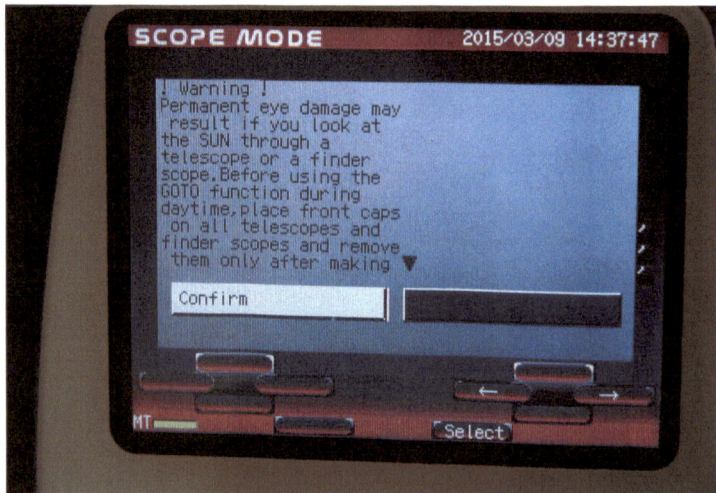

Fig. 7.2 Warning Displaying (Chen)

If observing at night, just select Confirm. If attempting daylight observation of the Sun, just select Confirm. By Not Confirm, the Star Book will not proceed forward. This is a Warning screen, probably there to prevent Vixen from legal actions against them when users attempt something stupid, such as pointing the telescope at the Sun without proper safeguards. **Observing the Sun should only be attempted when using solar filters or solar telescopes specifically designed to observe the Sun safely. Observing the Sun without proper filtering equipment can cause damage to your eyes** (Fig. 7.2).

The next screen is common to both the original Star Book and the Star Book TEN. There are perhaps more problems with setup from this particular screen than any other single screen during initialization procedures for the Star Book and Star Book TEN. The Home position for the beginning of the alignment procedure requires the telescope pointing to the West. As illustrated in the example given in the introduction of this book, if the mount and telescope is pointed to the East instead of the West, performing GoTo searches during the alignment procedure results in the telescope being aimed towards the ground.

Don't laugh. This seems to happen often. It is easy to see how this mistake occurs. Standing behind a polar aligned telescope facing North, the image on the Star Book display points the user's right. The user mistakenly then points the telescope to their right, unknowingly starting the alignment procedure by pointing to the East in establishing the Home position. A closer examination of the image on the display shows the picture is taken from the front of the telescope and mount with the image being taken in a South direction. With the telescope again pointing to the right, but by facing South the right hand direction actually points the telescope to the West.

When standing behind the telescope, facing North, the user needs to know that West is to their left and East is to their right. If standing in front of the telescope and facing South, West is to the right and East is to the left.

Unfortunately, when this mistake is made, the next screen on the original Star Book does not alert the user of this oops. The display will show a sky view map as if the mount has pointed the telescope West. Don't blame Vixen. The Home position identifies to the Star Book where the alignment begins, it's starting point. Up to that point, the Star Book firmware doesn't know where it is pointing. A review of other computerized GoTo systems shows the establishment of a Home position is a common requirement. Some systems require a North pointing Home position, while others share with Vixen the West pointing Home position. Fortunately, once experienced and corrected, this mistake rarely occurs again (Fig. 7.3a, b).

With the telescope now pointing in the proper West facing position, the process of selecting stars for alignment purposes begins.

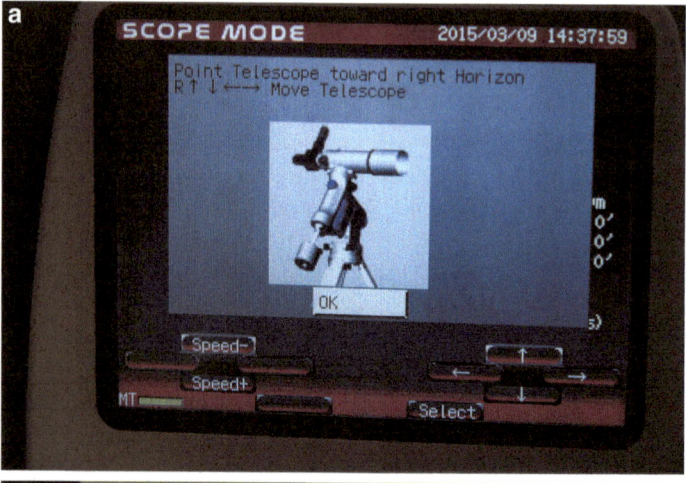

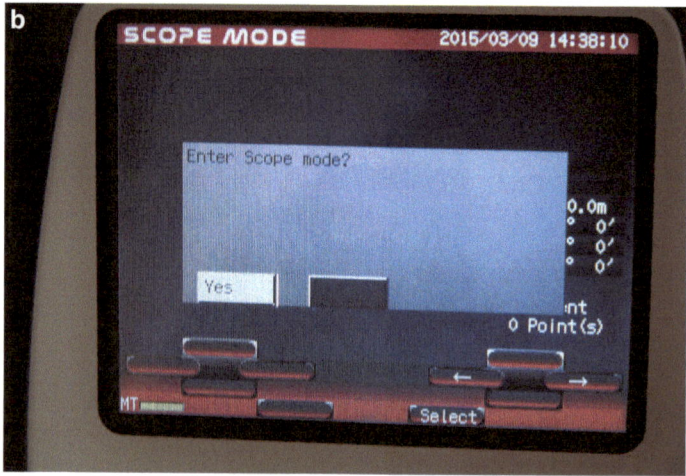

Fig. 7.3 (**a**) Point Telescope to West Screen (Chen). (**b**) Enter Scope Mode screen (Chen)

Pressing the right-hand selection key, graphically shown as "Chart" on the display, changes the display to Chart Mode (Fig. 7.4a, b).

By pressing the selection key, the Chart Mode screen pops up. Pressing the left hand left directional key reveals the Object menu. Most of this menu will be discussed later in this chapter. Of interest in the initializing process is the fifth line on the menu marked "Star". By highlighting and selecting Star accesses the menu of named stars stored in the Star Book memory (Fig. 7.5).

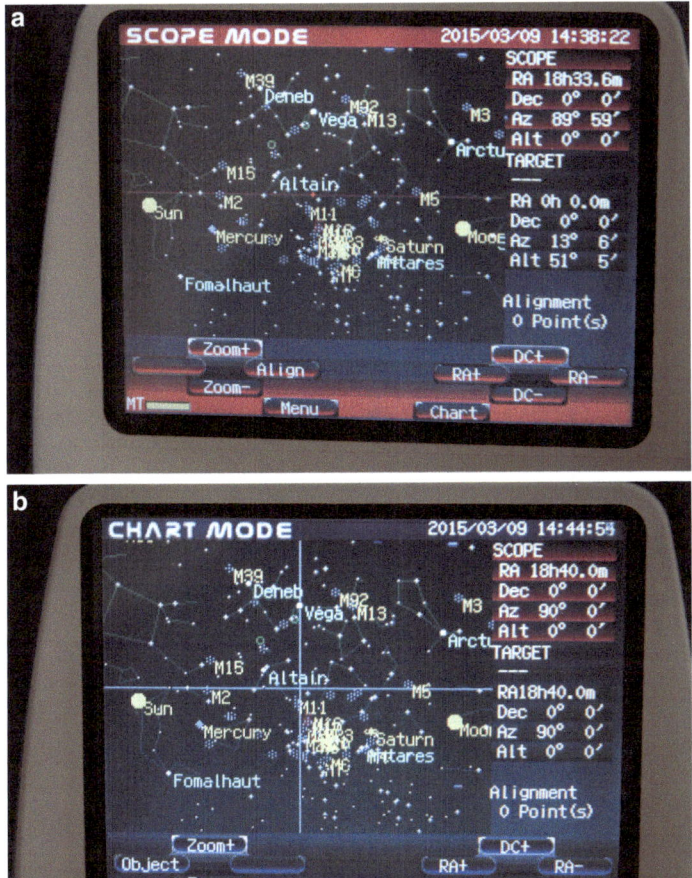

Fig. 7.4 (**a**) Scope Mode Screen following the Point the Telescope West screen (Chen). (**b**) Chart Mode Screen (Chen)

Unlike the Star Book TEN, which places a double circle next to the star name for those stars visible at the time and date, the original Star Book does not. The original Star Book firmware simply does not list those objects that are not visible at that time of night or time of the year.

Once the alignment star is centered in the eyepiece, hit the align key as indicated on the Scope Mode screen, and an alignment screen shall appear. Answer Yes and the Star Book will recognize the alignment star as a alignment data point.

The strategy of selecting stars for alignment is similar to the process outlined for the Star Book TEN. The Star Book firmware allows for up to six stars or objects to

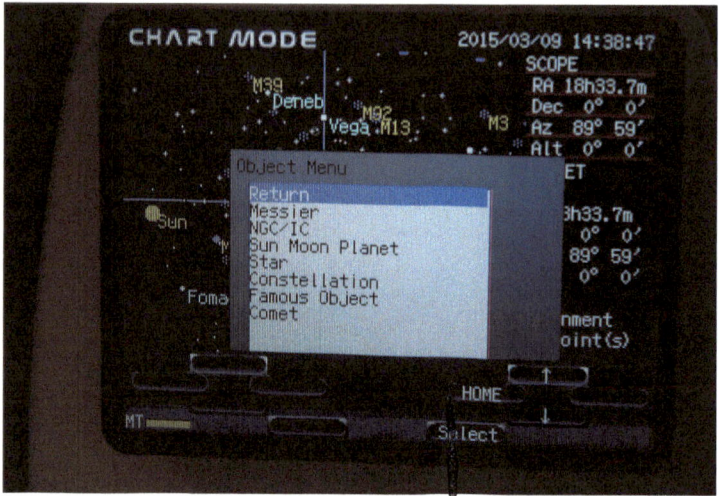

Fig. 7.5 Object Menu (Chen)

be used for alignment. The more stars used for alignment, the more accurate the GoTo search. A minimum of two stars are needed for the GoTo searched object to fall into the field of view of a low power wide field eyepiece (approximately 25× to 35× and 65° or greater AFOV). Try to select the first two stars wide apart and to either side of the meridian.

For example, during a winter time observing session, if a bright star in the constellation Orion is selected (such as Betelgeuse or Rigel), the second star selected might be Pollux or Castor in the constellation of Gemini, or Capella in the constellation Auriga.

During the summer months, a good pair of alignment stars might be Antares in the constellation Scorpio and Vega in the constellation of Lyra.

As the user, there are a number of good pairings to use from the list of bright stars from the Star Proper Name Menu for the initial two star alignment. The one restriction for selecting alignment stars is that the Star Book firmware requires a separation of at least 10° from each object. It is advisable to chose alignment stars at least 20° above the horizon. The refractive effects of the Earth's atmosphere near the horizon introduces inaccuracies into the Star Book firmware. Also avoid alignment stars at the zenith for a more practical reason. Using straight through finders or red dot finders cause the user to get into an unnatural and uncomfortable position when viewing straight overhead. Proper centering of the alignment star can be compromised.

If the observing is to be limited to a particular quadrant of the sky, it is possible to select two stars located in the same region so long as the 10° rule is followed. Returning to the Orion example, Rigel could be selected in Orion, and then Aldebaran in the constellation of Taurus would result in a good alignment pairing.

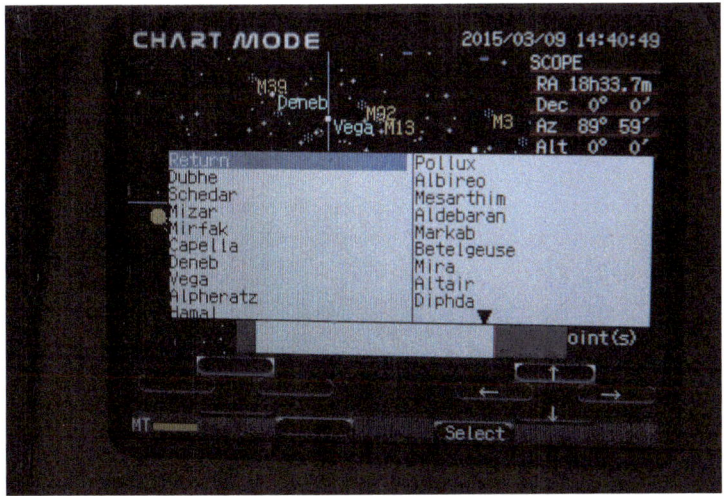

Fig. 7.6 Named Stars Menu (Chen)

In this example, objects such as the Crab Nebula Messier 1 and the Orion Nebula Messier 42 should be easily found with the eyepiece field of view, although not necessarily centered in the eyepiece. This quadrant based alignment will work for objects in the vicinity near Orion, but a search across the sky to opposite, for example an object in Ursa Major, will result in a miss (Fig. 7.6).

As additional stars are added to the list of alignment points, the Star Book becomes more accurate in its GoTo searches. Some users have reported that the threshold appears to be four to six alignment stars or objects are needed to attain center of the field searches. Let experience be your guide.

A major difference between the Star Book TEN and the original Star Book is the newer system allows for management of the alignment stars. This capability does not exist with the original Star Book. Care must be taken with each alignment star.

Vixen goes through great lengths in their instruction manual on using pinpoint star images as alignment objects, and not extended objects such as emission nebulas or open clusters. Since there is no definable center of an extended object, this is good advice. The exception are smaller planetary nebulas. Planetary nebulas, such the Ring Nebula Messier 57 or the Ghost of Jupiter NGC 3242, are small enough to be useful as alignment objects. So, by planning the observing session properly, any GoTo searches for small planetary nebulas can also be used for alignment purposes.

At the risk of sounding redundant from the previous Star Book TEN chapter, a little discussion bears repeating on the actual process of centering the selected star in the field of view of the eyepiece to establish alignment. Don't be surprised if the mount slewing to the first target star is so inaccurate that the selected star falls outside the field-of-view of the finderscope. It is not uncommon for the Star Book/Sphinx combination to miss the first alignment star by 5 or more degrees, well outside the field-of-view of a finderscope. The challenge then becomes using the slow motion controls of the Star Book to bring the target star into the view of the finderscope, and then centering the star in the main telescope eyepiece. Make sure that the star that is sighted in the crosshairs is the named star being used as an alignment star! Things tend to look different through the eyepiece of a finderscope, and users have been fooled into centering onto the wrong star in both the finderscope and the main tele-scope eyepiece. When this happens, the Star Book doesn't know any better (it's not artificial intelligence), and the alignment is off and all subsequent GoTo searches are in error and will point in the wrong point of the sky. A good method to use to avoid this problem is to select first magnitude or brighter stars for alignment.

Again, equipping your telescope with a red dot finder or a reflex finder, such as a Telrad or Rigel Quickfinder, will also be of aid. Using a red dot finder and select-ing first magnitude stars eases sighting the alignment, especially when the tele-scope and mount misses the targeted alignment star by several degrees. Often the first selected alignment star can be off center by several degrees. When this occurs, the miss aimed star can fall several degrees outside the field-of-view of the finder-scope, which can cause problems in correcting using the slow motion control direc-tion buttons on the Star Book. The view in the finderscope eyepiece can be disorienting, depending on the finderscope design (straight-thru or right angle viewing and erect or non-erect image). From actual hands-on experience, the red dot finder actually works better for alignment purposes. Remember, the goal of the finder is to get the alignment star into the field-of-view of the main telescope eye-piece, and is not the end-all and be-all of the alignment process. The alignment star can still be seen, even when it falls outside of the red dot display window, and using the directional buttons of the Star Book the process of bringing the target alignment star and the red dot becomes intuitive (Fig. 7.7).

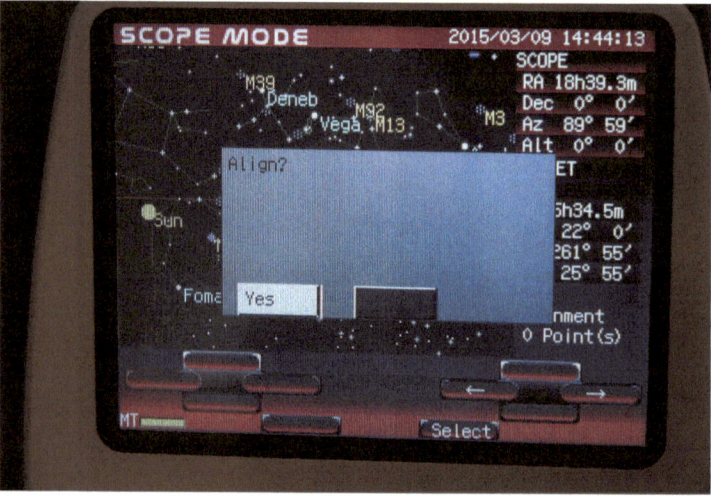

Fig. 7.7 Align screen (Chen)

If you find that the display backlight goes out, turning the display black, while you are performing alignments, the remedy is the following: In Scope Mode, hit the left hand key for Menu and scroll down the pull down menu to the backlight listing. Select Backlight and change the setting to 60 s or more to help with the alignment process.

Assuming the finderscope has been properly aligned, once the alignment star is in the crosshairs of the finderscope, the star should appear in the field-of-view of the eyepiece on the telescope. Slow the slewing rate down by using the left side Up button, which zooms in the object on the Star Book display while slowing down the slew rate. Center the star in the field of view, and for greater accuracy, change to a higher magnification and center the star again. Many users will use an eyepiece that will give them 100× magnification and is equipped with a reticle to center up the star. Once everything is centered, hit the align button, as indicated on the Scope Mode display. Then go onto the next alignment star.

Occasionally, users have been known to align their GoTo telescopes on the wrong star. The view in the finderscope can sometimes be disorienting. On occasion, users of the Star Book TEN, the original Star Book, and other competing GoTo systems have peered through the finderscope eyepiece and centered the crosshairs on the wrong star! This frequently happens with inverted views from the finderscope and using less than first magnitude stars. Second and third magnitude stars can present alignment problems, especially with similarly bright nearby stars in the neighborhood. For example, by mis-identifying and aligning on Pollux instead of Castor in the constellation Gemini, an extreme mis-alignment can occur and GoTo search problems will result. The best advice is take your time and don't rush the alignment process and make sure the correct star is sighted in for alignment.

It is recommended that following the setup routine that a few test GoTo searches be performed to assure the quality of the alignment. During the Winter, obvious objects such as M42 The Orion Nebula or M45 The Pleiades are good choices to make as test targets to make sure everything is working right. The Summer test objects might include easy objects, such as M13 the Great Globular in Hercules or M57 the Ring Nebula. The Autumn test targets might include NGC 869 and NGC 884 the Double Cluster in Perseus or M31 the Andromeda Galaxy. Spring objects might include Mizar in Ursa Major or M51 the Whirlpool Galaxy. The Star Book and Sphinx mount should locate these objects easily and with good accuracy when the setup and alignment procedures have been perform accurately. If the GoTo search for these objects result in a complete miss, check your inputs and alignment procedures, and at worst then apply the TOTOTA principle: Turn the system Off, Turn the system back On, and Try Again.

Experience has shown that even with care and following procedures to the letter, sometimes the alignment is off and erroneous GoTo searches occur. As with any computer program, sometimes for unknown reasons, the computer glitches. The Star Book is no exception. When this happens, TOTOTA. Switch off the mount and Star Book, turn it back on and try again. Use a different set of alignment stars the second go around and success should follow.

The alignment process becomes easier with practice. It is common for new users of the Star Book and the Sphinx mounts to initially spent up to an hour on their first setup and alignment procedure. After a few uses of the equipment, the controls and the setup will become familiar and routine, and the setup process becomes a 5 or 10 min process.

GoTo Search Operation

Now that the Sphinx mount is properly setup, the Star Book has been initialized and the alignment process has been completed, the fun begins.

There are basically two methods to perform a GoTo search using the Star Book. The first method is to use the stored pull-down lists of Messier objects, NGC and IC objects, solar system objects, named deep sky objects, and named stars. In general, this is the most popular method amongst owners of Star Book. An alternative method is to use the directional controls and place the crosshairs on the graphical display on an object of interest, zoom-in on the object to provide a more precise centering in the crosshairs, and then activate the GoTo search from the star map. Both search methods work equally well.

It should be noted that adding up the number of objects in the pull-down lists does not come close to the claimed 22,725 objects. The named lists include 109 Messier objects, 4980 NGC/IC objects, and the Sun, Moon, and planets of the solar system. Vixen claims 17,635 SAO catalogue stars are available, but not all are listed on the pull-down menu. The Star menu only lists named stars. The remaining stars are precisely plotted on the graphical star map display and can be searched and found by using the crosshair search method.

The side note from the Star Book TEN applies here: To those who bought naming rights to a star or stars for loved ones as a Christmas gift or birthday gift, these names are not recognized by the International Astronomical Union, or the scientific and astronomical communities. Although a cute and sometimes thoughtful gesture, the money spent on these naming stars is not official and is a pure ripoff of consumers. So don't expect to see your Uncle Floyd or Grandma Sally's name in the Star Book named star database. It won't be there.

By pressing the selection key, the Chart Mode screen will pop up. Using the left hand set of keys, pressing the left directional key reveals the Object menu. The first line after Return on the menu marked "Messier". By highlighting and selecting Messier accesses the menu of 109 Messier objects stored in the Star Book memory. A screen showing details of the selected object is displayed (Fig. 7.8a–d).

By moving the highlight down with the directional key, a selection can be made then by hitting the selection key. The screen will return to Chart Mode, with the star chart showing the location of the selected Messier object. Hit the GoTo key (the right hand side selection key), and a Confirm GoTo screen will then appear.

The Star Book is smart in that it will display only the Messier objects that are within viewing for the particular date and time the mount is being used (Fig. 7.8b).

By pressing the selection key, the Chart Mode screen will pop up. Using the left hand set of keys, pressing the left directional key reveals the Object menu. The second line after Return on the menu marked "NGC/IC". By highlighting and selecting NGC/IC accesses the menu of 4980 NGC and IC objects stored in the Star Book memory (Fig. 7.9).

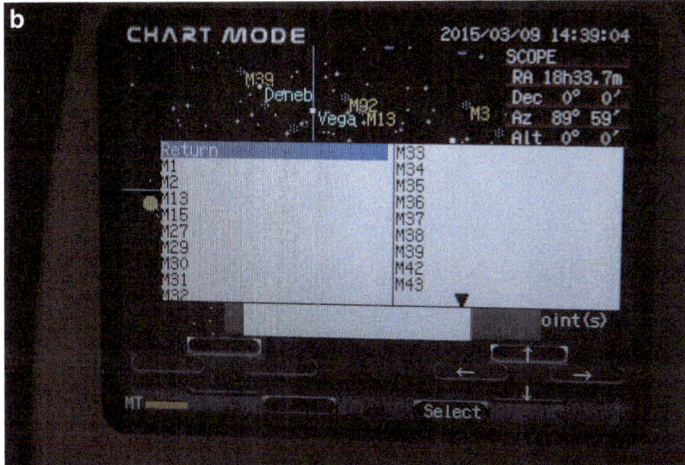

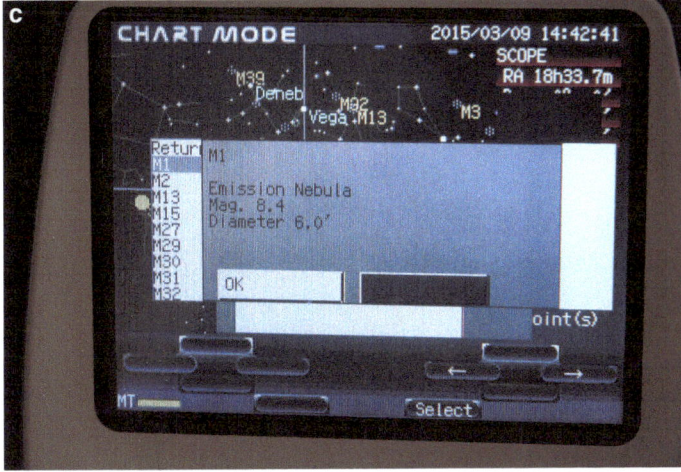

Fig. 7.8 (a) Objects Menu (Chen). (b) Select Messier Objects screen (Chen). (c) Selected Messier Object details (Chen). (d) Confirm screen (Chen)

Fig. 7.8 (continued)

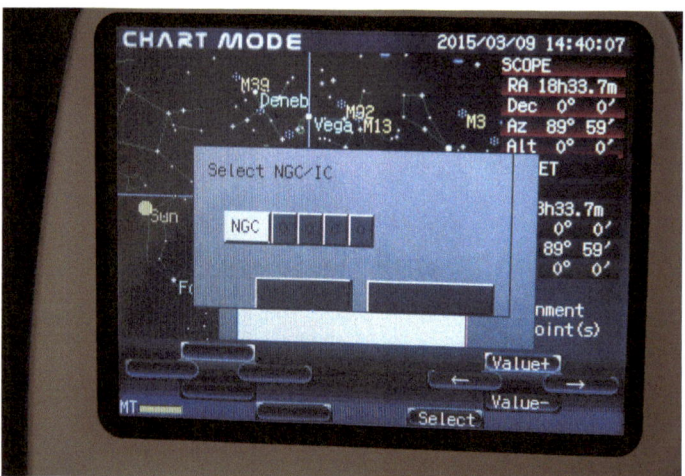

Fig. 7.9 Select NGC/IC screen (Chen)

By moving the highlight down with the directional key, a selection can be made then by hitting the selection key. A screen showing details of the selected object is displayed. The screen will return to Chart Mode, with the star chart showing the location of the selected NGC or IC object. Hit the GoTo key (the right hand side selection key), and a Confirm GoTo screen will then appear.

The Star Book is smart in that it will display only the NGC or IC objects that are within viewing for the particular date and time the mount is being used.

By pressing the selection key, the Chart Mode screen will pop up. Using the left hand set of keys, pressing the left directional key reveals the Object menu. The third

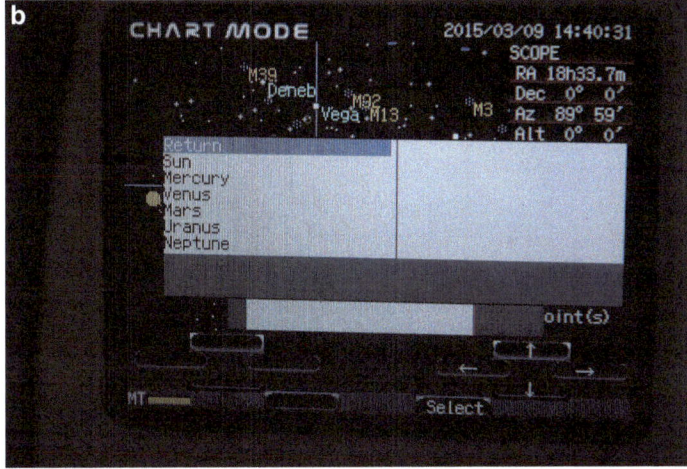

Fig. 7.10 (a) Objects Menu (Chen). (b) Sun Moon Planet screen (Chen)

line after Return on the menu marked "Sun Moon Planet". By highlighting and selecting Sun Moon Planet accesses the menu of the Sun, Moon, and the major planets stored in the Star Book memory. Since the Star Book's genesis in the early 1990s, Pluto has been categorized as a minor planet.

By moving the highlight down with the directional key, a selection can be made then by hitting the selection key. A screen showing details of the selected object is displayed. The screen will return to Chart Mode, with the star chart showing the location of the selected solar system object. Hit the GoTo key (the right hand side selection key), and a Confirm GoTo screen will then appear.

The Star Book is smart in that it will display only the solar system objects that are visible for the particular date and time the mount is being used (Fig. 7.10a, b).

By pressing the selection key, the Chart Mode screen will pop up. Using the left hand set of keys, pressing the left directional key reveals the Object menu. The fifth line after Return on the menu is marked "Constellation". By highlighting and selecting Constellation accesses the menu for all the 88 constellations in the Northern and Southern hemispheres stored in the Star Book memory.

By moving the highlight down with the directional key, a selection can be made then by hitting the selection key. A screen showing details of the selected object is displayed. The screen will return to Chart Mode. with the star chart showing the location of the selected constellation. Hit the GoTo key (the right hand side selection key), and a Confirm GoTo screen will then appear.

As with the Star Book TEN, this constellation database is useful when using Chart Mode and performing searches using the crosshair method (Fig. 7.11a, b).

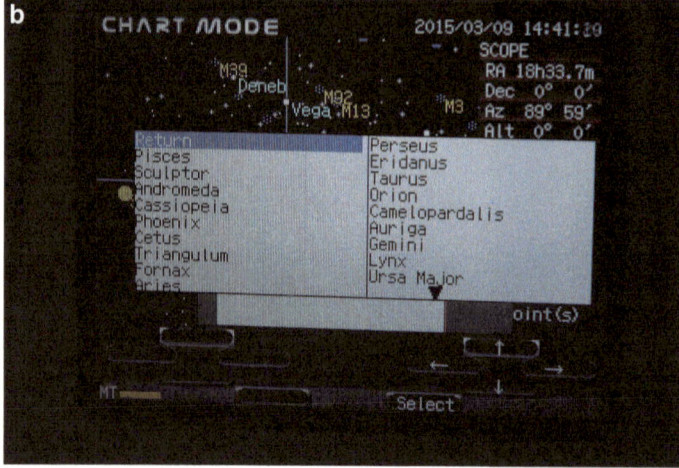

Fig. 7.11 (a) Objects Menu (Chen). (b) Select Constellation screen (Chen)

By pressing the selection key, the Chart Mode screen will pop up. Using the left hand set of keys, pressing the left directional key reveals the Object menu. The sixth line after Return on the menu is marked "Famous Objects". By highlighting and selecting Famous Objects accesses the menu for 25 of the most famous named deep sky objects stored in the Star Book memory. For example, the Famous Objects included are the Andromeda Galaxy, the Ring Nebula, the Blue Snowball Nebula, and the Double Cluster(s).

By moving the highlight down with the directional key, a selection can be made then by hitting the selection key. A screen showing details of the selected object is displayed. The screen will return to Chart Mode, with the star chart showing the location of the selected famous named object. Hit the GoTo key (the right hand side selection key), and a Confirm GoTo screen will then appear.

The Star Book is smart in that it will display only the Famous Objects that are within viewing for the particular date and time the mount is being used (Fig. 7.12a, b).

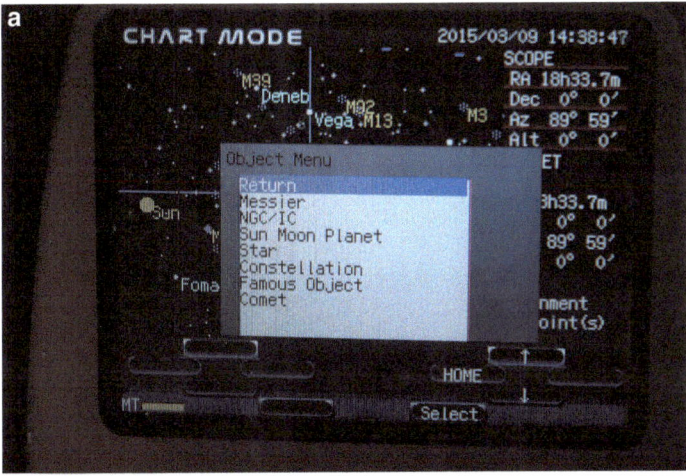

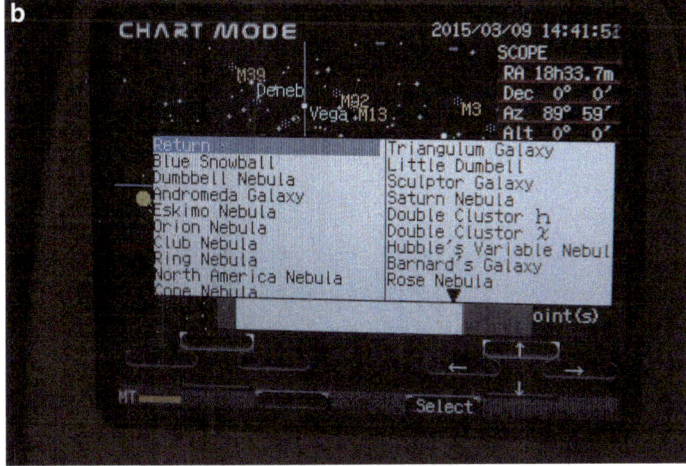

Fig. 7.12 (**a**) Objects Menu (Chen). (**b**) Select Famous Object (Chen)

By pressing the selection key, the Chart Mode screen will pop up. Using the left hand set of keys, pressing the left directional key reveals the Object menu. The last line after Return on the menu is marked "Comet". By highlighting and selecting Comet accesses the menu for any comet whose orbital elements have been downloaded into and stored in the Star Book memory. The ease of storing the orbital elements of comets is not like that of the Star Book TEN. The user must interface the Star Book with a PC computer in order to load the Star Book memory with data files for comets. The process of loading comets and their orbital elements is not very convenient, and an appendix at the end of this book for the process of adding comet orbital elements has been provided by Vixen. (The applicability and usefulness of the Vixen procedures has not been confirmed by this author, since the author uses an Apple computer.)

If the comet database has been established, the procedure for selection is as the previous object lists. By moving the highlight down with the directional key, a selection can be made then by hitting the selection key. The screen will return to Chart Mode, with a star chart showing the location of the selected comet. Hit the GoTo key (the right hand side selection key), and a Confirm GoTo screen will then appear. The Star Book and Sphinx mount will track the comet, now just merely compensate for the Earth's rotation (Fig. 7.13a, b).

The System Menu is accessed from the Scope Mode by selecting the Menu key on the left side key pad.

The System Menu allows the user to customize the operation of the Star Book to suit their personal needs (Fig. 7.14).

By selecting the first line item on the Systems Menu marked Chart Setting, a number of settings for the display of the star chart are accessed.

- Display Style allows the user to choose between an AltAz or RA/DEC depiction of the sky. The default setting is the AltAz setting.
- The Constellation setting gives the user the choice of having the constellations shown:
  - without lines,
  - with lines
- Constellation name can be displayed in long form, short abbreviated form, or not displayed.
- The Bayer Designation of stars can also be selected, with similar settings as the previous named stars. The default setting is Magnitude 4 stars or brighter.
- Confirm GoTo screen can be toggled either On or Off. The default setting is ON.
- Bright Stars can be selected as Always Off, Always On, Zoom1, Zoom2, or Zoom3. The default setting is Always On. The necessity for this system setting is a user preference. It is recommended to leave it in the default setting (Fig. 7.15).

By selecting the second line item on the Systems Menu marked LCD Adjust, the contrast and the brightness of the display of the star chart can be adjusted. LCD Adjust allows the user to dim the Star Book display to help preserve night vision, and to adjust the contrast of the display to make the display easier to read. The availability of this display control is valuable as ambient lighting conditions change during an evening of

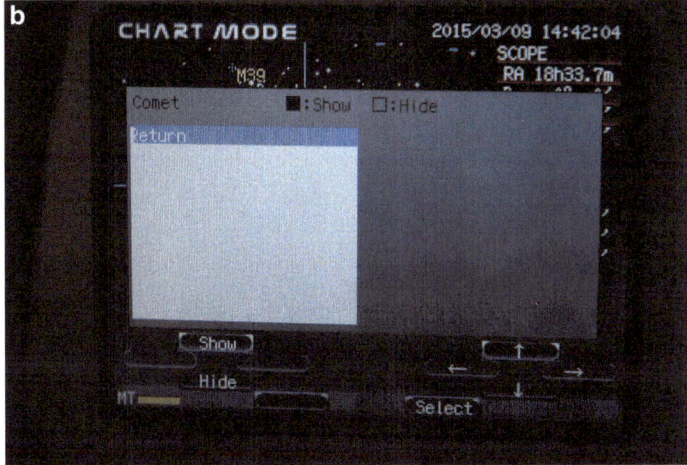

Fig. 7.13 (**a**) Object Menu screen (Chen) (**b**) Comet screen (Chen)

observing. For example, if a waxing crescent Moon is present early in the evening, the user will probably want to dim the display later as the Moon sets (Fig. 7.16).

The selection of the third line item on the Systems Menu marked Backlight, the time the display remains lit between actions can be adjusted. This action provides a level of battery power savings. During the alignment procedure, it is recommended that the backlight remain on for more than 60 s. This gives the operator plenty of time to perform the alignment procedure at a leisurely pace without the display screen blacking out and causing confusion (Fig. 7.17).

For those Sphinx mounts equipped with the polar axis alignment scope, the fourth line on the System menu allows the user to adjust the brightness of the reticle within the polar alignment scope (Fig. 7.18).

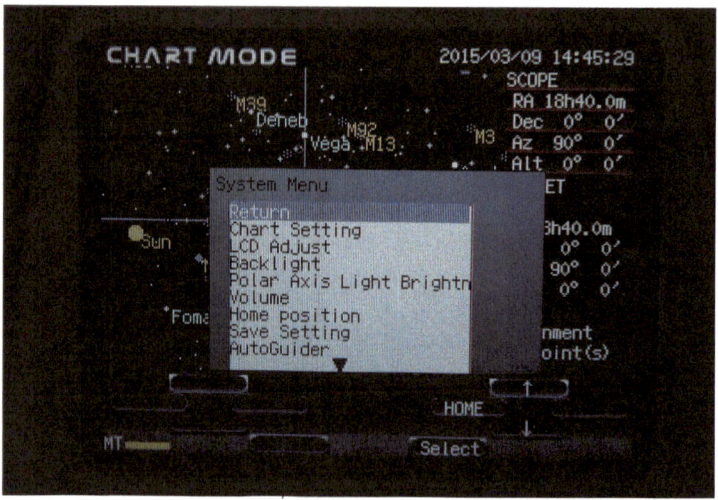

Fig. 7.14 System Menu (Chen)

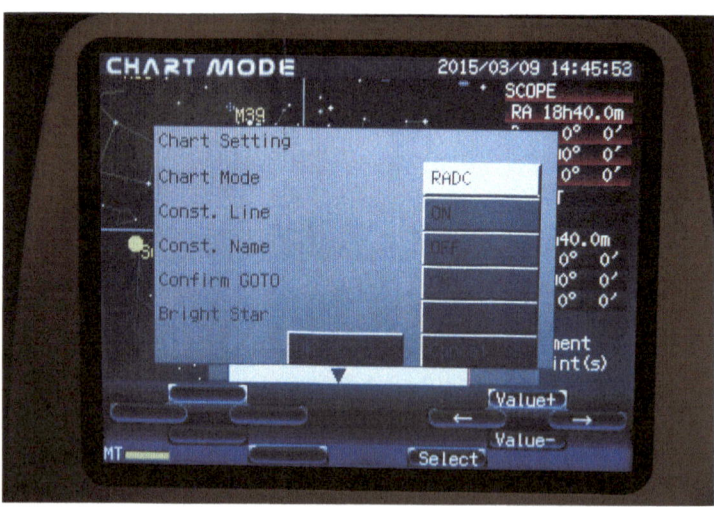

Fig. 7.15 Chart Setting screen (Chen)

The fifth selection on the System menu allows the user to adjust the loudness of the Star Book beep that announces the completion of a GoTo action. The sound volume is adjustable from 0 to 100 in increments of 1. The recommendation here is to leave the sound level at the default setting of 100 (Figs. 7.19 and 7.20).

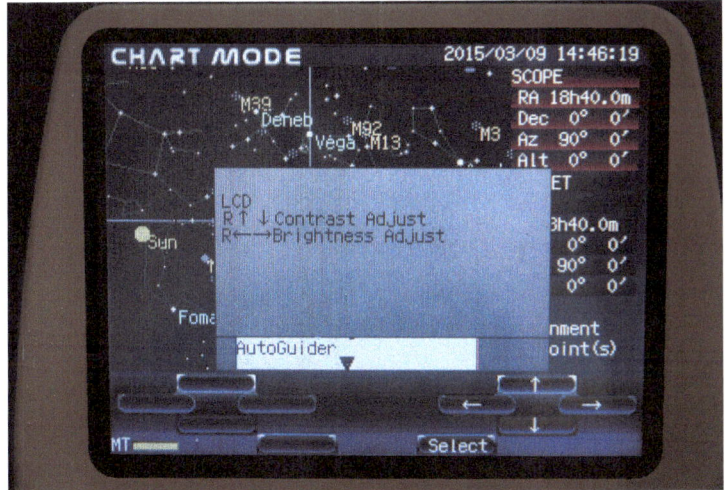

Fig. 7.16 LCD adjustment screen (Chen)

Fig. 7.17 Backlight adjustment screen (Chen)

By scrolling down the Systems Menu using the down directional arrow on the right side of the Star Book to the seventh line item, the Home Position is highlighted. Hitting the Select key selects Home Position and the mount will automatically slew the telescope to the West pointing home position used for the original alignment process. The usefulness of this particular command depends on the user. It is helpful for preparing for the end of an evening's observations. Or if the setup

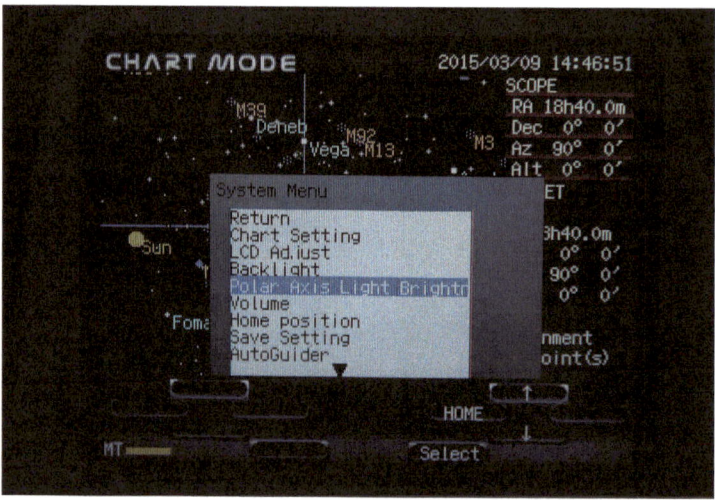

Fig. 7.18 System Menu - Polar Axis light brightness (Chen)

Fig. 7.19 System Menu - Volume adjustment (Chen)

routine needs to be re-initiated due to a power outage, poor initial alignment, or similar event (Figs. 7.21 and 7.22).

The selection of the eighth item is the Save Setting. This selection must be used to save all settings under the System Menu. If the settings are not saved, any changes made will apply only to the current session. When the Star Book is cycled

Fig. 7.20 Volume adjustment screen (Chen)

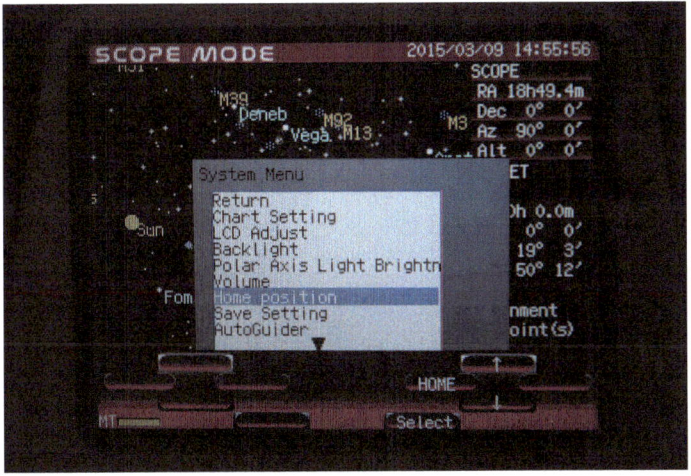

Fig. 7.21 System Menu - Home Position screen (Chen)

off, the settings would then return to the previously saved settings. For this command to work properly, all axis locks must be maintained from the previous use. Save Setting is best used in an observatory installation (Fig. 7.23a, b).

If an Autoguider is used, the Autoguider selection is used to determine the type of autoguider to be used, and the autoguider speed in right ascension and declination.

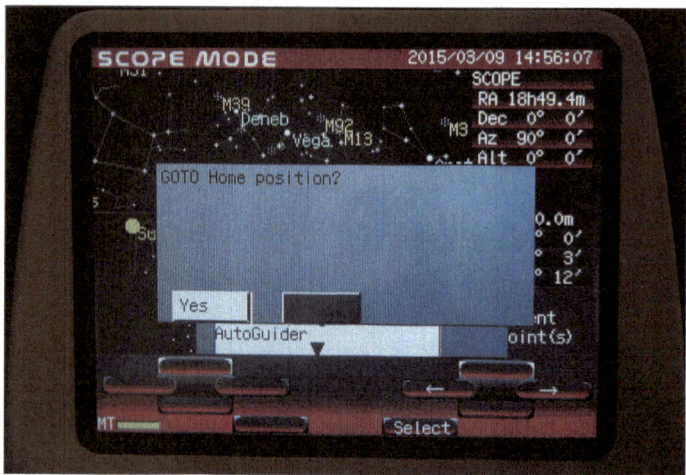

Fig. 7.22 GoTo Home Position screen (Chen)

The two selections available for autoguider type is AGA-1 and Standard. The range of speeds are 1–99 (Fig. 7.24).

PEC selection is used to initiate the recording of the periodic error of the drive system. This is only needed when astrophotography is being attempted. By recording the periodic error, the Star Book drive system can compensate for minute errors caused by the gears of the drive system. All mounts, not matter how well constructed, have minute errors due to manufacturing tolerances. PEC allows for electronic compensation of the periodic error.

By activating the PEC Start, the Star Book will record the minute errors in the gear drive system of the Sphinx, and according compensate during use thereafter (Figs. 7.25 and 7.26).

Backlash Compensation setting, with a range of 0–99, provides correction for mechanical inaccuracies of the gears and bearings of the Sphinx mount. The adjusts are both right ascension and declination separately. Again, all mounts, not matter how well constructed, have minute errors due to manufacturing tolerances that cause backlash. The Star Book features an electronic compensation so the backlash is no longer an issue (Figs. 7.27 and 7.28).

The next item on the System Menu is the adjustment of the GoTo speed. The Star Book firmware allows for two speed setting, either Middle or High. (Fig. 7.29)

Change GoTo Speed allows for the user setting of the slewing speed, with a speed range from either High to Middle (Fig. 7.30a, b). Star Book selection enables to the user to find the firmware version level of the Star Book. In the appendices are the Vixen

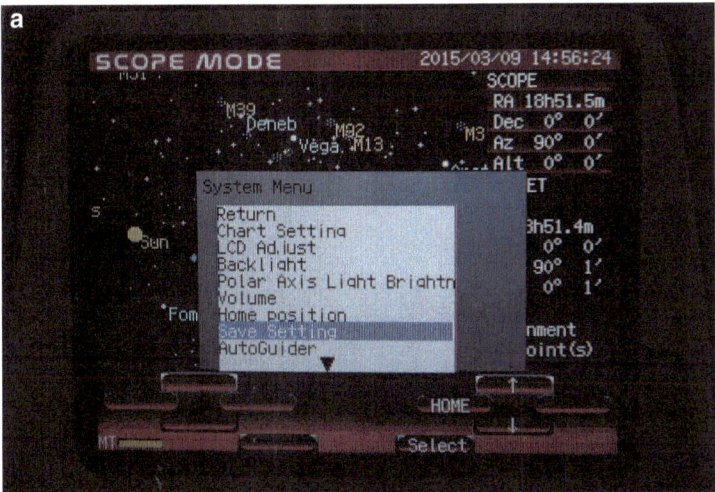

Fig. 7.23 (**a**) System Menu - Save Setting (Chen). (**b**) Save Setting Menu screen (Chen)

instructions for updating the firmware for both the Star Book TEN and the original Star Book. If a user is unfamiliar with downloading firmware updates, or is uncomfortable with the process, please see your Vixen dealer or contact Vixen directly and request their technicians to accomplish the update for you. The service may cost a little money, but the peace of mind that you have with someone else doing the work properly will be worth it.

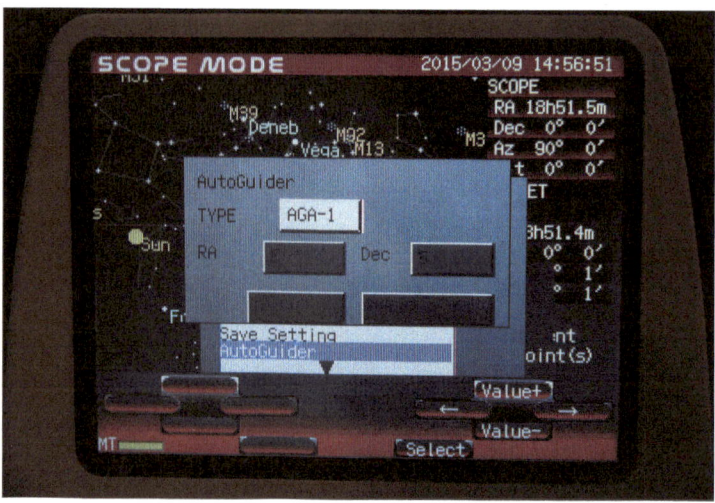

Fig. 7.24 Autoguider selection screen (Chen)

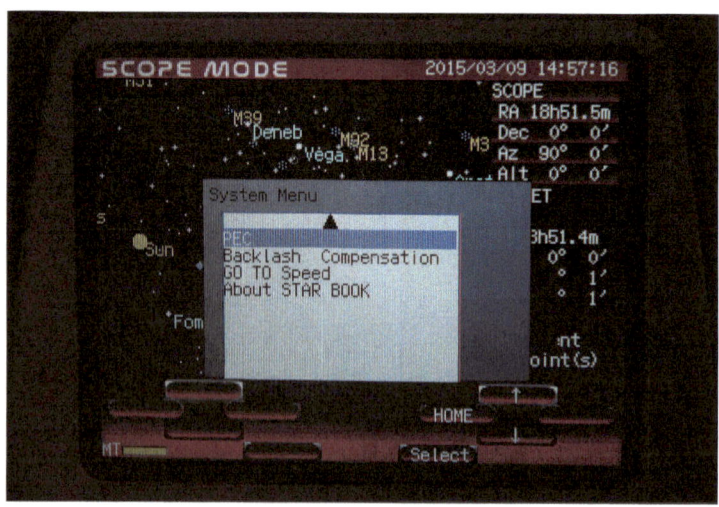

Fig. 7.25 System Menu - PEC screen (Chen)

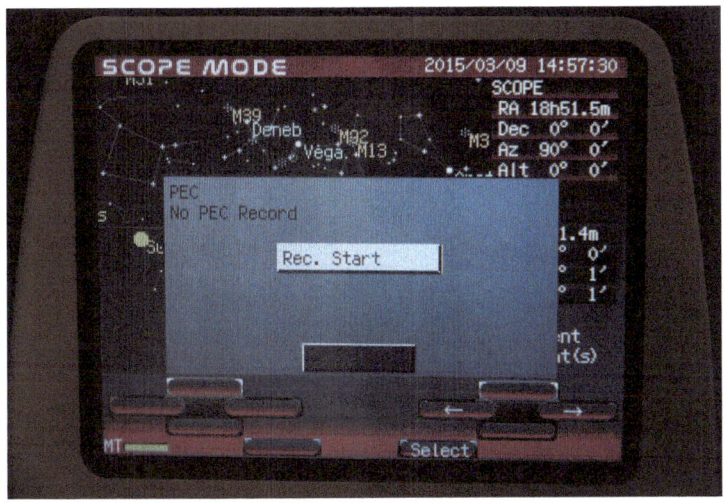

Fig. 7.26 PEC record screen (Chen)

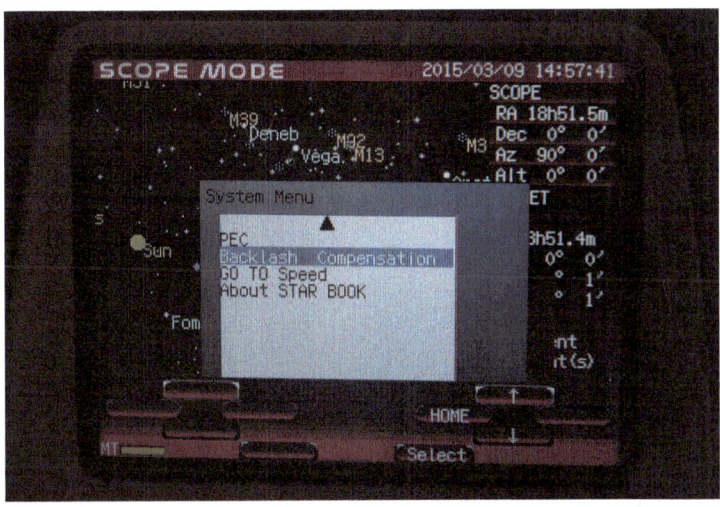

Fig. 7.27 System Menu - Backlash Compensation screen (Chen)

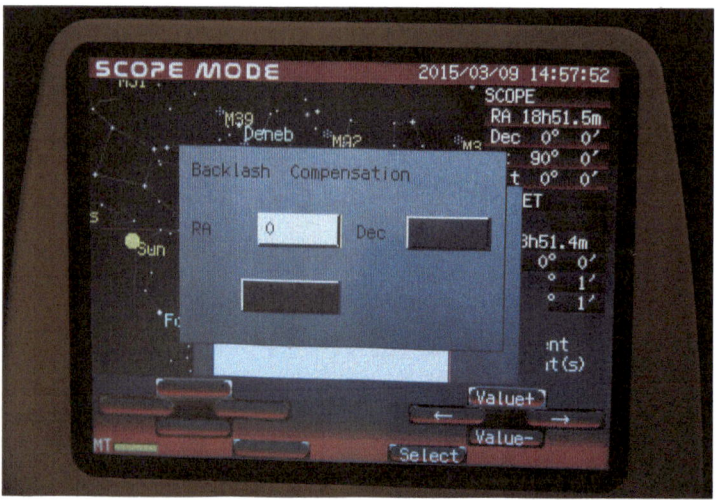

Fig. 7.28 Select Backlash Compensation Menu screen (Chen)

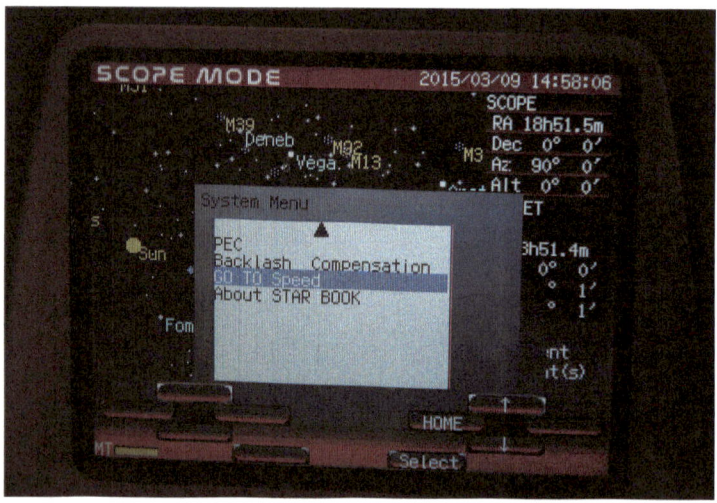

Fig. 7.29 System Menu - GoTo speed adjustment screen (Chen)

Fig. 7.30 GoTo Speed Menu screen (Chen)

Chapter 8

Learned Lessons
in Using the Original
Star Book

Experienced users of the original Star Book will adapt very quickly to the newer Star Book TEN system without any trouble at all. Just like their high technology cousins the personal computer or the ubiquitous smartphones, there are many commonalities shared between the first- and second-generation Star Book systems.

But each GoTo system has its quirks and idiosyncrasies that can prove challenging as the user climbs the learning curve. The original Star Book presents its own set of challenges that once the user has gained experience, the operation of the system becomes second nature.

General Comments on Using the Sphinx SXW Mount and Star Book

First and foremost, the Sphinx mounts and the Star Book requires a good source of electrical power feeding into the system. Whether using a Vixen power supply plugged into a wall outlet or a suitable battery pack, the proper voltage and amperage must be supplied to the mount and the Star Book to function properly. The low battery problem of the first generation Star Book systems becomes evident with the slowing of the DC servo motors and the inability to maintain tracking. Low battery power has no effect on StarBook controller functionality. Once the user recognizes that the Sphinx mount is starting to "drag its feet", the solution is easy. Either plug in a fully charged battery pack, or connect a wall power supply to the mount. Turn on the power switch, and mount and Star Book should power up correctly. There may be the necessity to proceed through the setup again, but the mount and Star Book will function correctly.

© Springer International Publishing Switzerland 2016   99
J.L. Chen, A. Chen, *The Vixen Star Book User Guide*, The Patrick Moore
Practical Astronomy Series, DOI 10.1007/978-3-319-21593-8_8

These are the general setup tips that apply to a Star Book equipped SXW or SXD mount:

- Use the most accurate latitude and longitude coordinates for your location. Modern smart phones are equipped with GPS capabilities that can provide the needed lat-long coordinates. Other sources for accurate lat-long information are car satellite navigation devices, such as Garmin and Tom-Tom, and applications on tablet devices.
- Cell phones and GPS devices are also good sources for accurate time settings.
- The telescope mount tripod should be relatively level. Precision is not necessary.
- It is best for accurate GoTo searches to setup as close to True North as possible. During the setup, Polaris the North Star, is your friend.
- Adjust the Backlight setting to 60 s or greater to help during the alignment process. The lowest setting available is 20 s, which from experience will interfere with the performance of the setup routine.
- Choose alignment stars in different parts of the sky, and at least 10° apart. The Star Book TEN allows for up to 20 alignment stars. The more the better, although a minimum of two are needed for initial relatively accurate searches. The Star Book system GoTo searches improve considerably with four alignment stars.
- During GoTo searches, be sure to use a low power wide field eyepiece. A 32–40 mm eyepiece with an apparent field of view of 65° or wider will insure the GoTo search will be successful, with the searched object visible within the field of view.
- The periodic error correction function is only used for astrophotography. For visual use, this function can be ignored, and PEC plays no part in the search accuracy of the Star Book.

Prior to setting up your equipment, make sure of the following:

- Make sure the Star Book and the power source is solidly connected to the Sphinx mount. Many quirky and erratic operations of the Star Book and the Sphinx mount can be traced to loose connections or dirty contacts. A cotton Q-Tip and denatured alcohol can be used to clean grubby contacts with no harm to the equipment and cable connectors.
- Make sure the red dot finder or finder scope is properly aligned prior to entering the alignment procedures. Life is easier with aligned finders.
- During the alignment procedure, lock both axises firmly and never loosen them during your session. All movement of the telescope is done using the electric drive motors of the mount.
- Insure that all internal batteries and external power sources are fully charged, fresh, and properly operating.
- Use the manufacturer provided cables. The back rooms of many telescope stores have examples of telescopes and their mounts with malfunctioning electronic and GoTo drives because owners have chosen to use off-brand cables, wiring, and power packs.

- Protect the Star Book controller from the elements and from dropping on the ground. This is a commercial grade electronic product, and not a military-spec or NASA grade device. There is no conformal coating on the circuit boards to protect from excessive moisture. The Star Book controller does not contain shock mount components. Don't abuse the equipment, and your equipment will stay away from the telescope backroom awaiting repair.
- If possible, use an inclinometer to set the latitude tilt of the mount. Here again, the handy smartphone will have an application available to use. Place the smartphone on the right ascension axis of the mount and adjust the tilt of the axis to equal the latitude where you will be observing.

Trouble Shooting the Star Book

A number of user problems with the Star Book have their roots in following correctly the setup data inputs and procedures. The most common symptoms of a setup error occurs when a GoTo command causes the scope to point downward at the Earth, or displaying the dreaded "Object is below the horizon" error message when the GoTo object is clearly above the horizon.

The following is a checklist of possible causes of the downward pointing telescope, inaccurate GoTo slewing, or the dreaded error message:

- Home Position screen. Make sure the telescope is pointing West. Then hit OK. If not pointed to the West, the Star Book electronics will cause the telescope to point to the ground!
- Substitute a red dot finder or a reflex finder for the finder scope. The initial setup GoTo to major stars tends to miss greatly and falls outside the field-of-view of a typical finder scope. A wider viewing angle is provided by the red dot finder with 1× magnification. 1× magnification will not bring other stars in the view, so the user can zero in on the known bright star, thus yielding a quicker setup time.
- Enter the current location with longitude (east or west) first, and then latitude (north or south). This is backwards for the normal "lat-long" order of coordinates. If the coordinates are input into the Star Book in the wrong order, the Sphinx mount and the Star Book reacts in a sometimes puzzling and sometimes funny manner. GoTo searches will be performed through a good part of the sky, misleading the user that all is functioning well. But parts of the sky will not be accessible. The telescope will either point to the ground, or the dreaded "Object is below the horizon" error message will be displayed on the Star Book. A good indication that the coordinates have been entered in the wrong order will be the Star Book establishing a horizon too high or too low in Chart mode.
- Pay attention to the local time setting and especially to the time difference from Greenwich time. Remember a change from Standard time to Daylight savings time results in a one hour change in the time differential between local time and Greenwich time. Also, the + or − sign for Greenwich Mean Time has a great

impact. If the mount is used east of Greenwich, UK, such as Europe or Asia, a "+" sign is used. If the mount is used west of Greenwich, UK, such as North America or South America, a "−" sign is used.

- A good CR2032 battery in the StarBook controller is needed to maintain the computer memory of longitude-latitude and time. A weak or dead battery will result in having to input time and location parameters every time the Star Book and the Sphinx mount is powered up.
- Check local time. Occasionally, the internal clock of the Star Book Ten loses time. This can occur even with a fresh battery. The wrong local time can be the cause of inaccurate GoTo slewing and downward pointing.
- During the initial setup, pick two widely spaced alignment stars in the east, and two widely spaced alignment stars in the west. The GoTo search will be relatively accurate at this point, with the searched object falling within the field of view of a low power wide field eyepiece. As the observing evening continues through the night, the deep sky objects that the observer has found using the GoTo can then be centered in the eyepiece field and used as additional alignment objects to improve the GoTo accuracy.
- Chose alignment stars at least 20° above the horizon. The refractive effects of the Earth's atmosphere near the horizon introduces inaccuracies into the Star Book firmware.
- Avoid alignment stars at the zenith. Using straight-through finderscopes or red dot finders cause the user to get into an unnatural and uncomfortable position when viewing straight overhead. Proper centering of the alignment star can be compromised.
- A major difference between the Star Book TEN and the original Star Book is the newer system allows for management of the alignment stars. This capability does not exist with the original Star Book. Care must be taken with each alignment star. If an error is recognized in one of the alignment stars, additional alignments must be made to cycle the older erroneous data out of the memory. Or use the TOTOA principle and start the alignment process over again.
- Experience has shown that even with care and following procedures to the letter, sometimes the alignment is off and erroneous GoTo searches occur. As with any computer program, sometimes for unknown reasons, the computer glitches, and the Star Book is no exception. The TOTOTA principle introduced in Chap. 5 applies again: Turn Off Turn On Try Again. When glitches occur, switch off the mount and Star Book, turn it back on and try again. Use a different set of alignment stars the second go around and success should follow.
- Be aware the official Vixen specification of the operating temperature for the Star Book is −20 to +40 °C, or −4 to +104 °F. At the lower temperatures, the Star Book will exhibit a non-responsiveness to commands. The lubricant in the Sphinx mount also tends to thicken at extreme low temperature and cause wear and stress on components on the Sphinx mount. It is recommended that the Sphinx mount and the Star Book not be used in sub-zero Fahrenheit temperatures. Extreme triple digit Fahrenheit high temperatures should also be avoided.

Many users complained about the Star Book display being too bright even on the lowest brightness settings. Vixen developed an update contained in firmware 1.0 build 18 to allow the brightness settings to be saved in memory. Vixen currently lists in its catalog an after market set of transmission reduction films which can be stuck to the screen to further reduce the brightness. Do-it-yourself brightness solution is to use neutral density film. A neutral density film is preferred rather than red tinted film as colored film will greatly reduce the readability of the color display. An alternate solution is suggested on-line from Peter Enzerink's enzerink.net site as:

> One user has also added a 3 m privacy filter. As described by the maker, this filter uses patented micro-louver technology which works like tiny vertical blinds. This restricts the viewing angle such that most of the light is directed directly outwards from the Star Book screen with much less lateral light escaping.
>
> You may wish to experiment on whether the louvers should be vertically or horizontally oriented. The filter also needs to be cut to size and one filter should be enough to cover more than one Star Book display which may help defer the cost.

Chapter 9

Notes on the Star Book-One and Star Book-S Systems

Over the years, Vixen has produced and marketed hand controllers under the Star Book title that are subset variances to the Star Book TEN and the original Star Book.

Star Book One

In 2014, Vixen introduced the Star Book One dual axis handheld controller. It is best to think of the Star Book One as a Star Book TEN without the GoTo capability, large size star chart display, and celestial object database. Using a Star Book One allows the use of the Sphinx family of mounts as conventional dual axis drive equatorial mounts.

 Since the introduction of the Sphinx SXW in 2004, Vixen found itself manufacturing and supporting two dissimilar product lines of equatorial mount: the Great Polaris series and the Sphinx series. In general, the Great Polaris series of German equatorial mounts were aimed at the telescope owner who did not want computer automation and just required single or dual axis motor drives with slow motion axis control via a hand controller. As stated numerous times heretofore, the Sphinx series represented the high tech computer automation segment of the marketplace. In a practical business and manufacturing move, the long-lived Great Polaris series of mounts were allowed to fade into amateur astronomy history. The new business model that Vixen adopted is to manufacture a single line of Sphinx mounts that would work with either the high-tech Star Book TEN controller or the lower-tech Star Book One controller. The Sphinx mount can now work with a choice of controllers, giving the customers the flexibility to buy a fully computerized mount, a lower more basic capability and upgrading to a Star Book TEN at a later date, or a basic control capability for those who have no need for GoTo capabilities (Fig. 9.1).

© Springer International Publishing Switzerland 2016
J.L. Chen, A. Chen, *The Vixen Star Book User Guide*, The Patrick Moore
Practical Astronomy Series, DOI 10.1007/978-3-319-21593-8_9

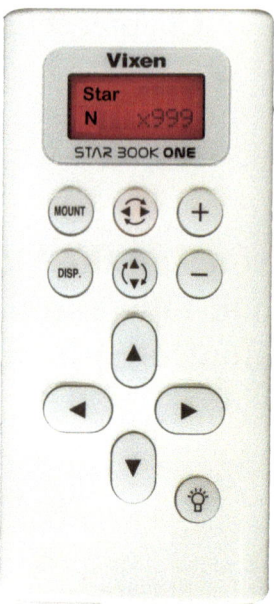

Fig. 9.1 The Star Book One (Vixen)

The Star Book One is very intuitive in its operation. The directional arrows provides slow motion control of the right ascension and declination axises. The + and − buttons allows for the selection of different preset four speed slewing speeds: 0.5×, 1.0×, 30×, and 999× (Fig. 9.2).

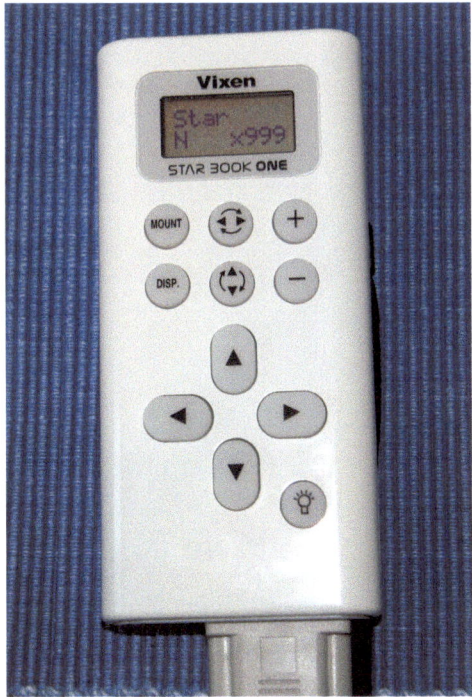

Fig. 9.2 The Star Book One with slewing speed at 999× (Chen)

The Star Book One allows for backlash compensation, periodic error correction (PEC), and use of an autoguider to achieve accurate tracking. To access these settings, the Mount button (or key) is pressed. The Track speed is displayed. The Speed selection can be made by pressing either the up or down directional buttons, allowing the user to select between Star, King, Lunar, Solar, Star at 0.1×, or Stop tracking.

Pressing the Mount button also allows the selection of Northern Hemisphere operation or Southern Hemisphere operation. This can be accomplished by first pressing the right directional button to scroll the menu to TrackDir (for track direction). North or South Hemisphere can be toggled by using the up or down directional button (Fig. 9.3).

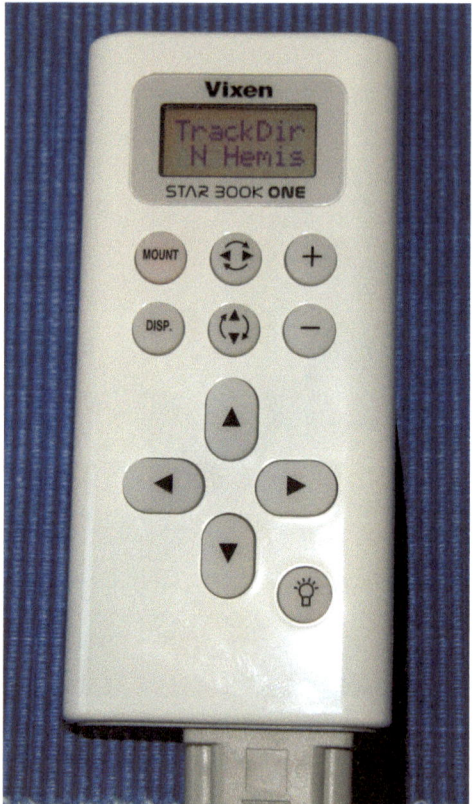

Fig. 9.3 The Star Book One for hemisphere selection (Chen)

By pressing the Mount button, and then pressing the right directional button twice allows access to the motor power setting, with a range of 1–4. Again the Up and Down directional buttons are used to select the lowest power usage setting of 1 to the highest power setting of 4. The three setting is the default setting. This power setting is used in order to conserve power from external battery packs (Fig. 9.4).

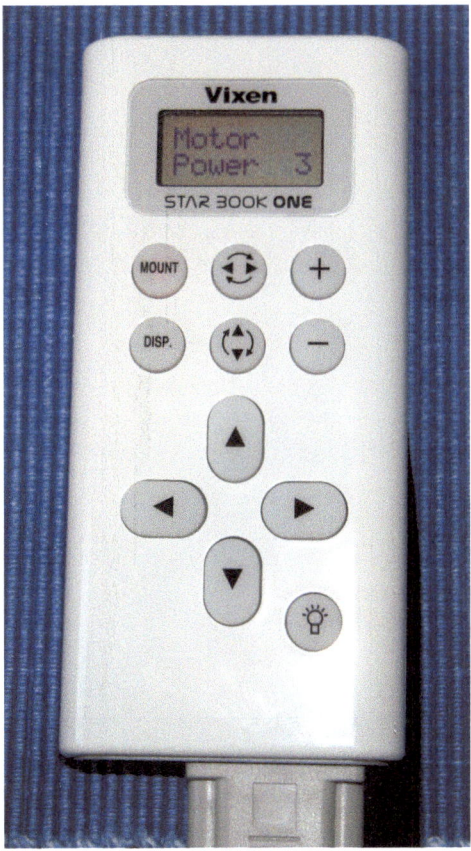

Fig. 9.4 The Star Book One motor power setting (Chen)

Changing the slewing speed can also be accomplished by using the Mount button. Stepping through the setting using the right directional key accesses the Slew to change from the preset four speeds to a continual variable speed setting. The range of slewing speeds can then be user adjustable from the 0.5× to the maximum of 999× (Fig. 9.5).

Backlash setting is then accessed with another right directional key push. The backlash then can be adjusted with the Up and Down buttons for the × direction (meaning one axis). One more press of the right direction key then addresses the Y direction backlash setting, adjusted again with the Up and Down keys (Figs. 9.6 and 9.7).

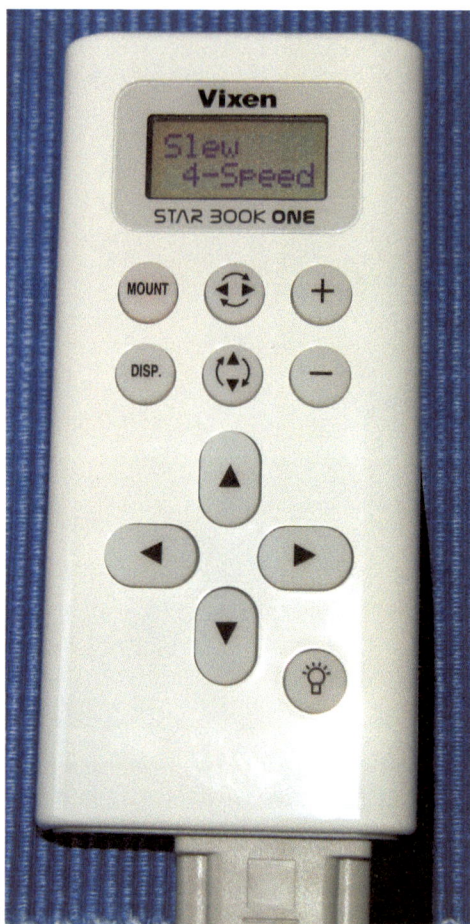

Fig. 9.5 The Star Book One slew rate (Chen)

Fig. 9.6 The Star Book One backlash compensation for X-axis (Chen)

Fig. 9.7 The Star Book One backlash compensation for Y-axis (Chen)

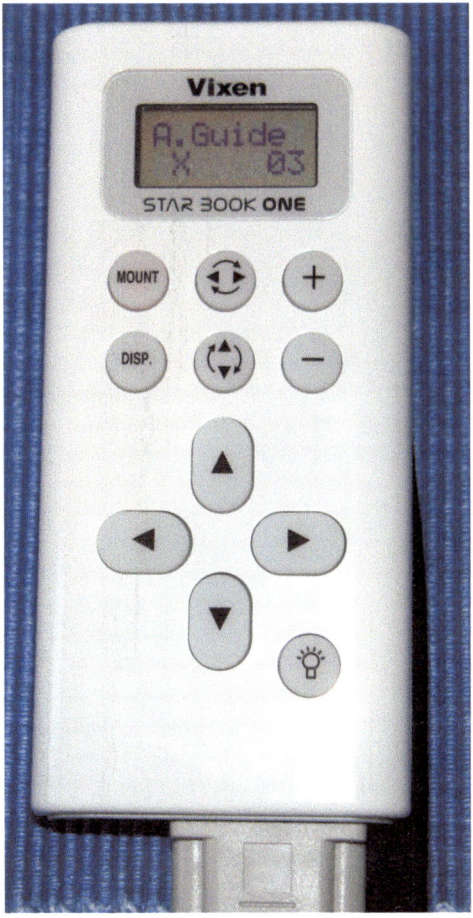

Fig. 9.8 The Star Book One auto-guider setting X-axis (Chen)

When performing astrophotography and using an auto-guider, the autoguider cable is connected to the base of the Star Book One. Then by pressing the Mount button and right scrolling until the display shows A. Guide, the user can select the autoguider slewing speed for X and Y directions. Again, the X and Y settings are separate and requires a right scrolling action via the right directional button (Figs. 9.8 and 9.9).

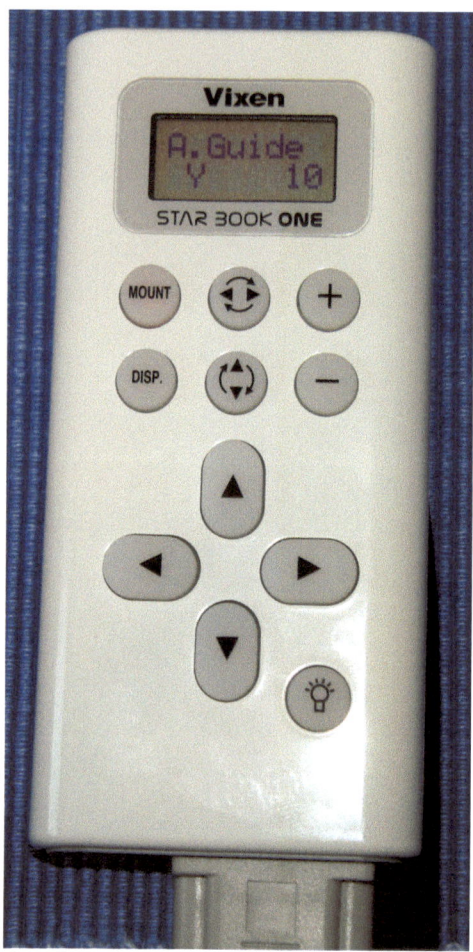

Fig. 9.9 The Star Book One auto-guider setting Y-axis (Chen)

The final Mount setting is the recording of the Periodic Error Correction, or PEC, data to compensate for any slight variation in the drive gearing. Press the Up or Down button to be prompted to begin recording. The + and − buttons will flash, and pressing the—key will begin the recording of the data. The Star Book One will initiate a 400 second countdown while it performs the PEC function (Fig. 9.10).

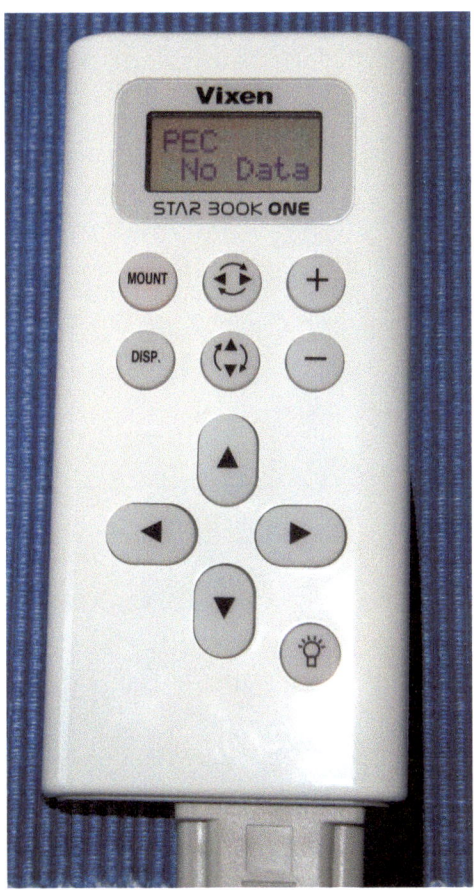

Fig. 9.10 The Star Book One PEC setting (Chen)

The Star Book One works in both northern and southern hemispheres. Tracking options available include sidereal rate, Kings rate, lunar rate, solar rate, and various speeds for time lapse photography. Access to these functions is gained through the pressing of the Mount button.

The button with the light bulb (see Fig. 9.1) activates a red LED light on the end of the Star Book One controller for use in reading star maps, manuals, or notebooks without ruining night vision (Fig. 9.11).

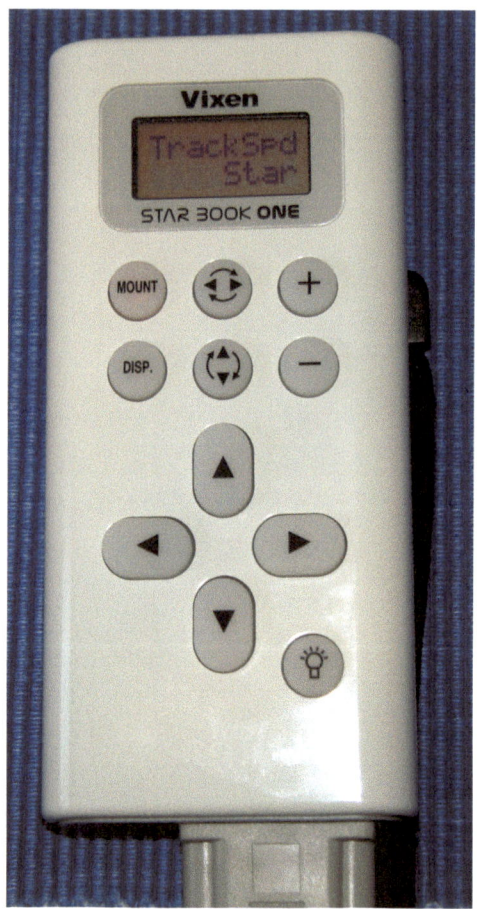

Fig. 9.11 The Star Book One tracking rate setting (Chen)

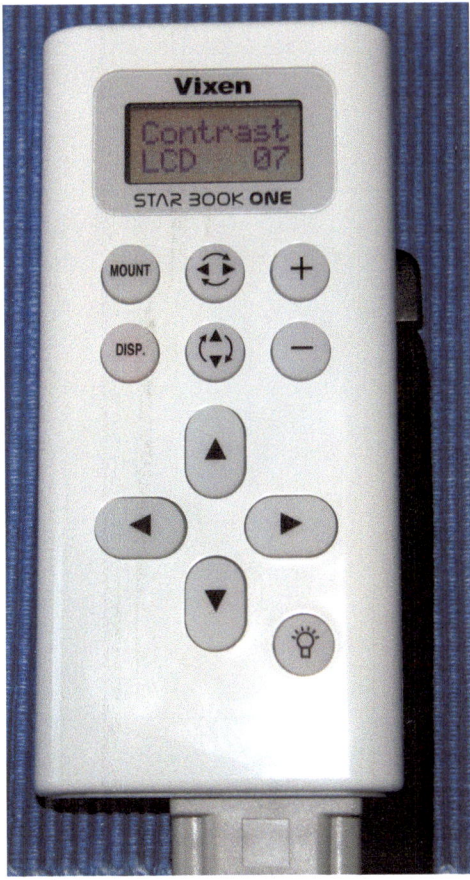

Fig. 9.12 The Star Book One display contrast setting (Chen)

Pressing the DISP button allows for the increase or decrease of the illumination and contrast of the Star Book One display (Figs. 9.12 and 9.13).

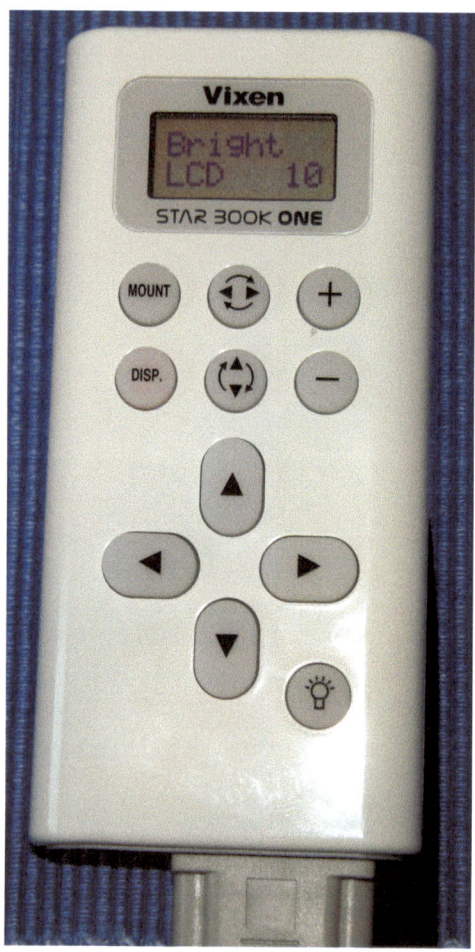

Fig. 9.13 The Star Book One display brightness setting (Chen)

Pressing the DISP button allows for the increase or decrease of the brightness of the Star Book One keys (Fig. 9.14).

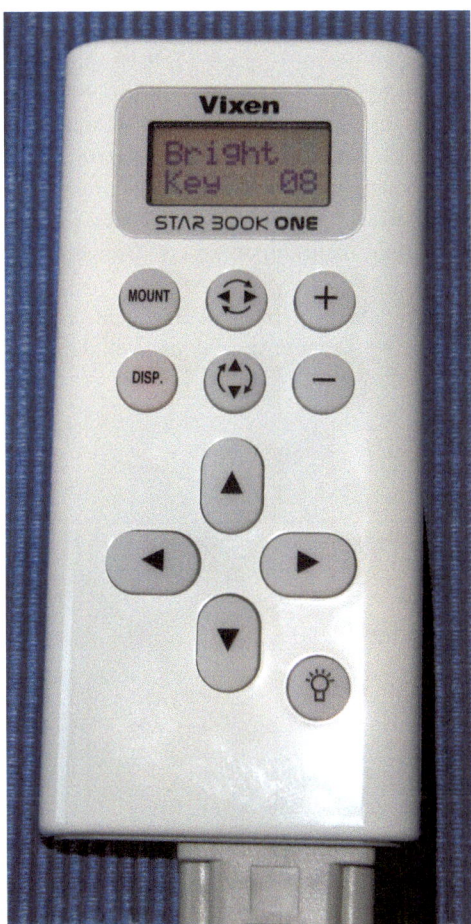

Fig. 9.14 The Star Book One key brightness setting (Chen)

Star Book-S

Vixen developed the Star Book-S following its successful introduction of the Sphinx SXW, SXD, and the first generation Star Book. The Star Book-S was a monochrome LCD display version of the Star Book with reduced dimensions , but the same functionality. It was an optional kit that added to the GP, GP2, and GPD2 mounts with the installation of suitable motors, and the controller came standard as part of the portable Skypod mount (Figs. 9.15, 9.16 and 9.17).

The Star Book-S is designed with a 2 in. monochrome LCD screen, and is functionally operated exactly like its large full-color screen equipped Star Book brother. Both controllers utilize the same firmware and operationally work the same, with the exception of the control keys which are located on the right side of the Star Book-S. The monochrome screen provided a courser resolution than its larger brother, and the user's right-hand thumb received a workout in operating the Star Book-S.

However, the operation of the Star Book-S is identical with the Star Book.

The Star Book-S is now discontinued, along with the GP, GP2, GPD2, and the Skypod mounts.

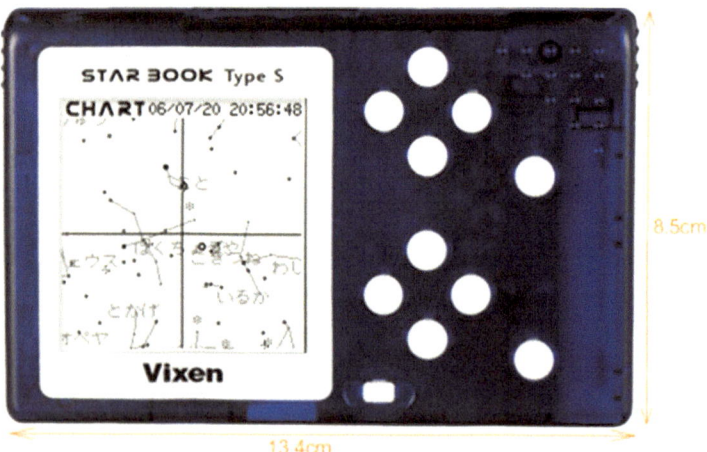

Fig. 9.15 The Star Book-S controller (Vixen)

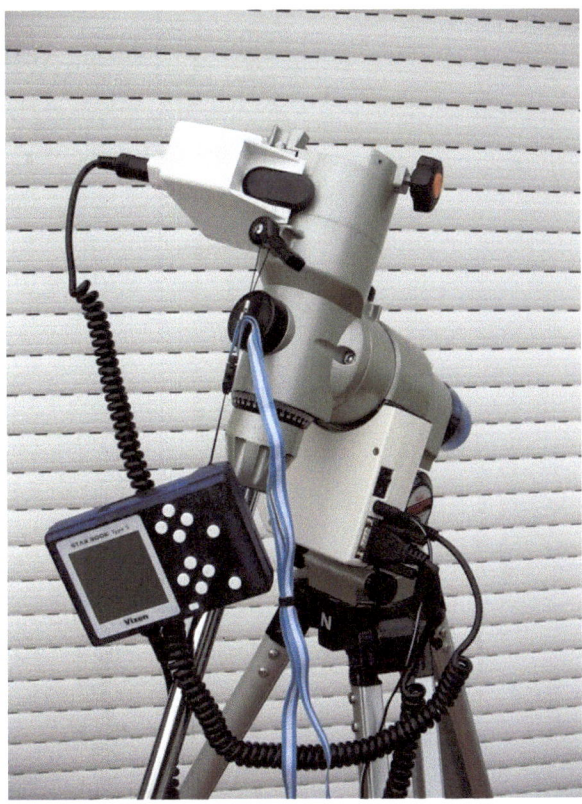

Fig. 9.16 The Star Book-S option installed on a GP2 mount (hands-on-optics photo archive)

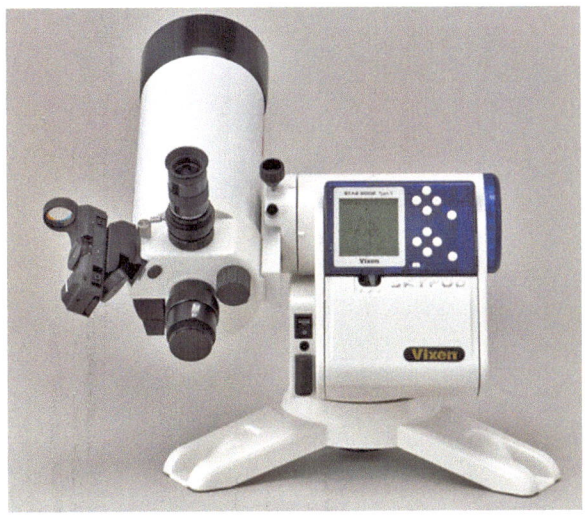

Fig. 9.17 The Skypod mount (Vixen)

Chapter 10

Accessories for Vixen Sphinx Mounts

Figure 10.1 shows the range of accessories available for the Sphinx family of mounts. There are variations dependent on which model is used. A number of the accessories are for visual or casual use of the Sphinx mount, and a number of the accessories are specific to astrophotography. Some accessories are attractive to users performing both visual and astro-imaging applications.

The current Vixen product lineup, which includes the SX2, SXD2, and SXP, offers each model of mount, with a choice of either with the Star Book One controller for basic use, or the Star Book TEN option for GoTo capability. The Vixen AXD is equipped with Star Book TEN as standard equipment.

The older original Sphinx mounts, which are the SXW, SXD, and AXD, came equipped with the original Star Book controller only. There are no basic controllers available for the original mounts. The original Star Book can operate the Sphinx mounts in a basic mode by not performing the alignment procedure. Using the direction keys, the user can slew to the desired target, and the right ascension motor will track the target.

Note: Neither the Star Book One nor the Star Book TEN will work with the older Sphinx mounts. The servo DC motors and the internal mount electronics are incompatible with the new technology that is encompassed in the Star Book TEN compatible mounts.

In reviewing the Sphinx accessory schematic, it should be noted a number of parts that are standard on each mount (Fig. 10.1). Items such as the counterweight

© Springer International Publishing Switzerland 2016   123
J.L. Chen, A. Chen, *The Vixen Star Book User Guide*, The Patrick Moore
Practical Astronomy Series, DOI 10.1007/978-3-319-21593-8_10

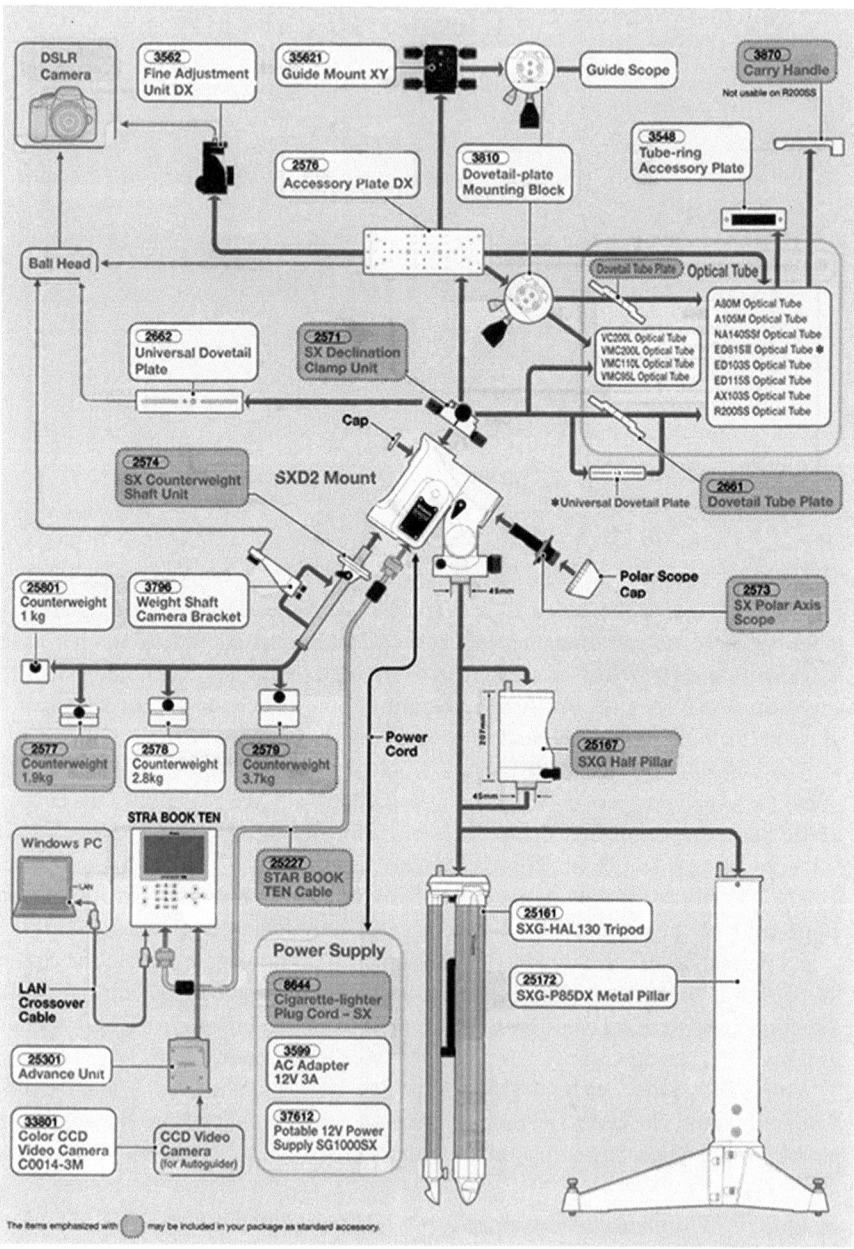

Fig. 10.1   The Sphinx mount family of accessories (Vixen)

shaft, polar scope end caps, the SX declination clamp unit, counterweights, and Star Book cable are included with each SX2, SXD2, and SXP mount. The polar scope used for polar alignment is standard with the SXD2 and SXP mount, and is offered as an option with the SX2 mount (Fig. 10.2).

Fig. 10.2 Polar scope for polar aligning the Sphinx mounts (Vixen)

Each Vixen Sphinx mount comes with at least one counterweight (Fig. 10.3). Depending on the size and weight of the telescope mounted, additional counterweights are available from Vixen, in 1.9, 3.7, or 8.15 kg increments (Fig. 10.3).

Fig. 10.3 Vixen counterweight (Vixen)

The SX2, SXD2, and SXP mounts are normally supplied with the HAL-130 tripod, a very stable and highly portable tripod (Figs. 10.4 and 10.5). As an option, the P85DX pillar is available for owners seeking additional stability, particularly when using the Sphinx on paved or concrete surfaces. For field use, it is advised to stick with the HAL-130 tripod.

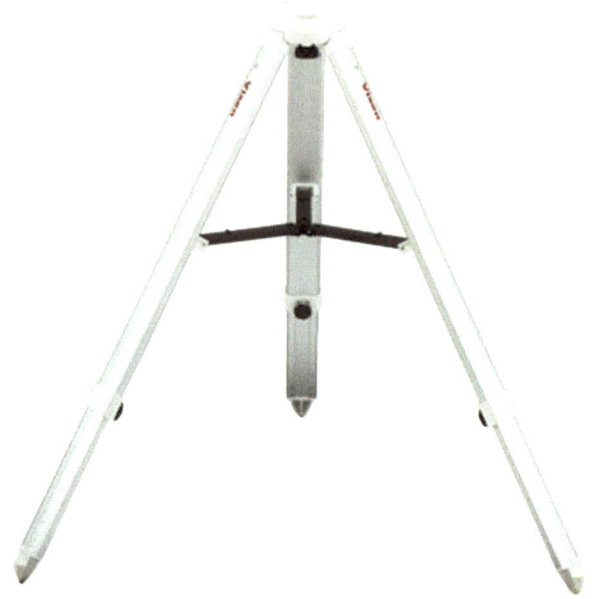

Fig. 10.4 Vixen HAL-130 tripod (Vixen)

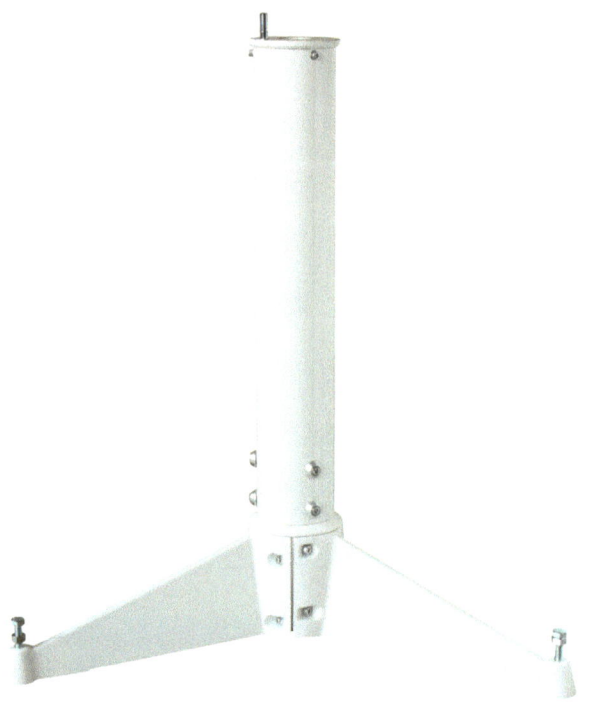

Fig. 10.5 Vixen P85DX metal pillar (Vixen)

An interesting and recommended option is adding the half pillar to the tripod mounted Sphinx mount (Fig. 10.6). This option is not needed when using a short tube Cassegrain, Schmidt-Cassegrain, Maksutov-Cassegrain, or similar compact telescope. However, when mounting a longer tubed refractor or Newtonian telescope, there is the very real possibility of the telescope tube striking one of the HAL-130 tripod legs during GoTo slewing. The addition of the half pillar raises the telescope and Sphinx mount high enough to eliminate the possibility of damage from the tripod interference during a slewing action. The half pillar also has the advantage of raising the height of the telescope and Sphinx mount to sometimes eliminate the need to extend the tripod legs to obtain a comfortable eyepiece position. Stability of the telescope and mount is also enhanced by not extending the tripod legs (Fig. 10.6).

Fig. 10.6 Optional half pillar (Vixen)

The Advance Unit (Fig. 10.7) for the Star Book Ten enables and enhances the system's auto-guiding capabilities and additionally allows to the viewing of a captured image from a CCD camera, record or play back from an SD card, and adjust the shutter exposure controls of a DSLR camera. The Advanced Unit plugs into the Star Book Ten by removing the expansion slot cover and sliding it into the slot (Fig. 10.7).

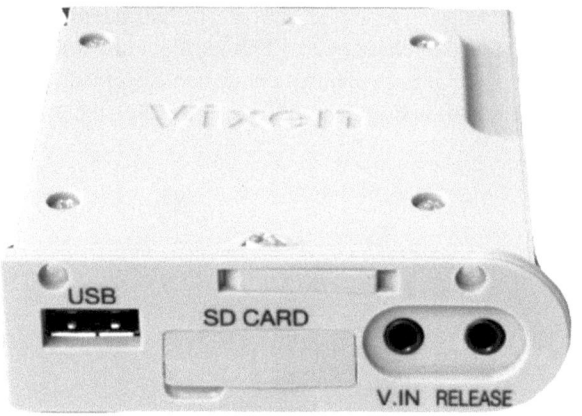

Fig. 10.7 The advance unit (Vixen)

The Advance Unit specifications are:

Signal type	NTSC
Image format	JPEG files less than 1 MB
Video format	AVI (Motion-JPEG) file: VGA, 15 frames per second Max 10 min of recording time, or until the memory card is full
Display resolution	VGA: 640×480 QVGA: 320×240
Zoom	0.5–4× digital zoom, QVGA only
Memory	SD/HC card slot, supports cards up to 32 GB
Connectors	Remote shutter release cable port 3.5 mm video-in port
Power requirement	Receives power from Star Book Ten
Operating temperature	32–104 °F (0–40 °C)
Dimensions	3.5×3.0×0.9″ (9.0×7.6×2.4 cm)
Weight	2.6 oz (74 g)

The Advance Unit works like a built in autoguider in combination with a CCD video camera. Images by AV analog can be displayed on the STAR BOOK TEN screen. Installing the Advance Unit on the STAR BOOK TEN enhances the auto-guiding capabilities. This allows the viewing of an image from a CCD camera, record or play back from an SD card, and the adjustment of the shutter exposure controls of a DSLR camera.

There are two modes of auto-guiding with the STAR BOOK TEN. One is the Advance Unit and another is using a commercially available external autoguider. The autoguider port is equipped with the STAR BOOK TEN for this purpose. The CCD video camera attached to the guide scope is pointed to the guide star. The signals from the CCD video camera make the autoguider work to automatically correct the drifting star. Highly accurate tracking is achieved and you are able to perform long period exposure astrophotography.

It is important to remember, the Advance Unit works only with the Star Book TEN. Older SXW, SXD, and ATLUX mounts equipped with the original Star Book design cannot utilize the Advance Unit.

A unique requirement that necessitated an easy solution for the original Star Book is the brightness film (Fig. 10.8). Many early adopters of the Sphinx mount and Star Book controller found that the illumination control of the original Star Book did not provide sufficient dimming of the display without affecting their night vision. For those who need additional dimming of the display, Vixen offers a self-adhering film that when applied to the Star Book display provides additional lowering of the illumination. The film is not permanent, and can be removed easily if necessary (Fig. 10.8).

Fig. 10.8 Star book brightness film (Vixen)

The subject of providing power to the Sphinx mount and the Star Book TEN or Star Book controller should be the highest priority of a Sphinx mount owner (Figs. 10.9, 10.10, and 10.11). As described earlier in this book, the mount and computer controller needs good power in order to function properly. The recommended solutions are as follows:

1. If AC is available, use the Vixen AC-to-DC adapter power supply. Yes, the Vixen power supply is pricey, but it definitely works with the Vixen mounts without question and there is potentially less risk of damaging the electronics. Off-brand power supplies will work, but great care must be made in their selection.

Fig. 10.9 Vixen power cable for car accessory sockets, portable car battery, etc. (Vixen)

Fig. 10.10 Vixen AC-to-DC adapter power supply (Vixen)

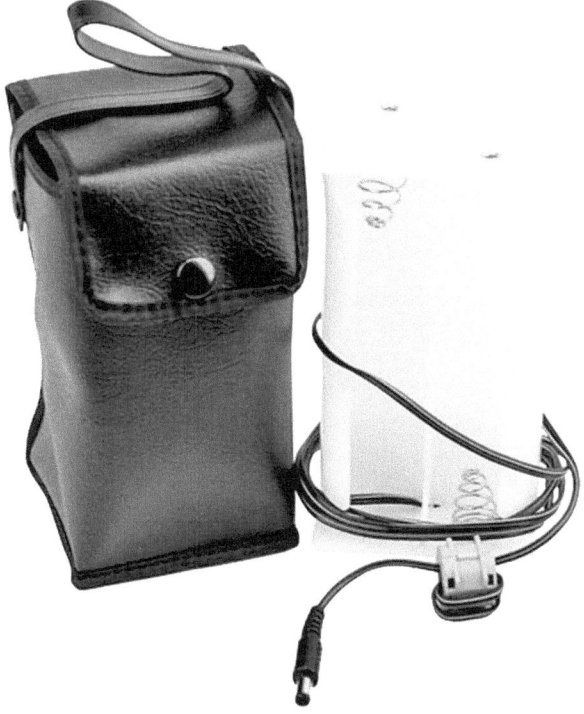

Fig. 10.11 Vixen battery box (Vixen)

2. For field work, obtain the Vixen power cable adapter and a high quality portable car battery power supply or Celestron's Power Tank. These portable battery power supplies will provide good DC power over the course of one long night, or possibly two or three shorter observing sessions. Here again, don't go cheap.
3. The Vixen battery box is marginal, at best, for field use. With heavy use of GoTo searches, the power provided by the battery box's eight D-cell batteries will only provide 1 or maybe 2 hours of reliable power to the mount and controller. The battery box might be a good source of backup power in case of extreme use resulting in a drained field battery pack.

This is the classic Vixen dovetail mounting bracket, designed to easily mount or dismount telescopes onto Vixen SX Series Mounts (Fig. 10.12). With Vixen Tube Rings or third party rings attached with 1/4-20 threaded holes to the dovetail, telescopes can be mounted and dismounted easily using the Vixen dovetail system. Other devices such as a camera can also be attached. Interchangeable with the SX tube plate. Two 1/4″-20 bolts are included.

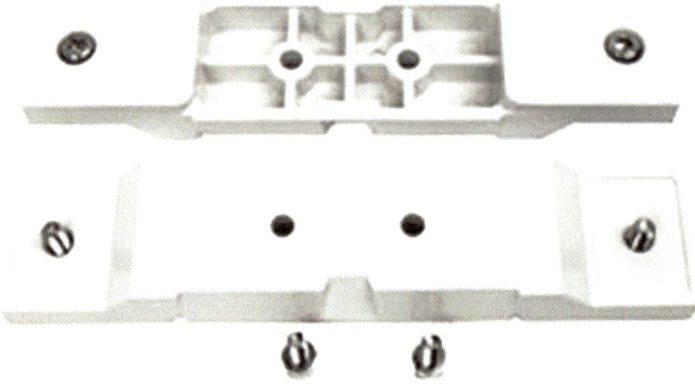

Fig. 10.12 Vixen dovetail plate (Vixen)

Allows attachment of camera to the 20 mm weight shaft found on Vixen Sphinx mounts (Fig. 10.13). Because the cameras weight is now helping counterbalance the Optical Tube, users are greatly appreciative of the fact they do not have to add MORE counterweights. Some use this to balance their Optical Tubes with Heavier cameras. Very multi-purpose.

Fig. 10.13 Weight shaft camera bracket (Vixen)

This Triangle Accessory tray provides a convenient place to set many of the things an amateur astronomer may be using through the observing night, such as eyepieces, charts, red flashlight etc (Fig. 10.14). While not only providing a temporary storage function, this tray also increases the overall stability of the tripod and reduces any vibrations seen through the telescope.

Fig. 10.14 Triangle accessory tray (Vixen)

This is the Deluxe Version of the Accessory Plate with over 32 mounting holes to fit almost every telescope mounting plate currently in production (Fig. 10.15). By sliding this adaptor plate into the standard Vixen Dovetail Mount, other telescopes and mounts can be mounted side by side.

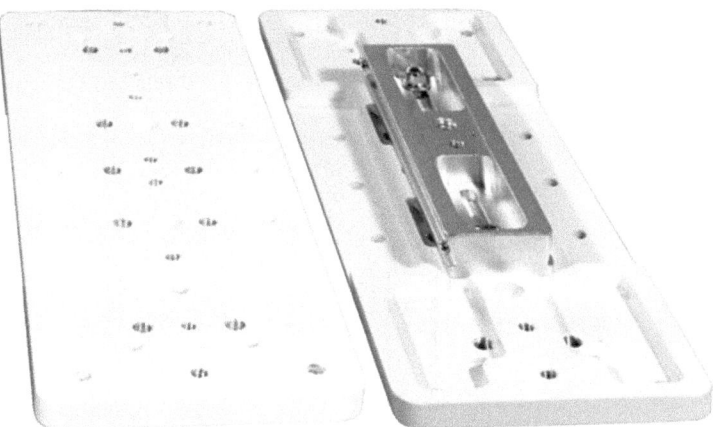

Fig. 10.15 Accessory plate DX (Vixen)

Aluminum case designed to securely carry the complete Sphinx SX mount, including Sphinx, Star Book, counterweights and battery case (Fig. 10.16, 10.17, and 10.18).

The tripod case is designed to carry the HAL-130 tripod.

And finally, a case is available to securely carry the Star Book TEN or original Star Book.

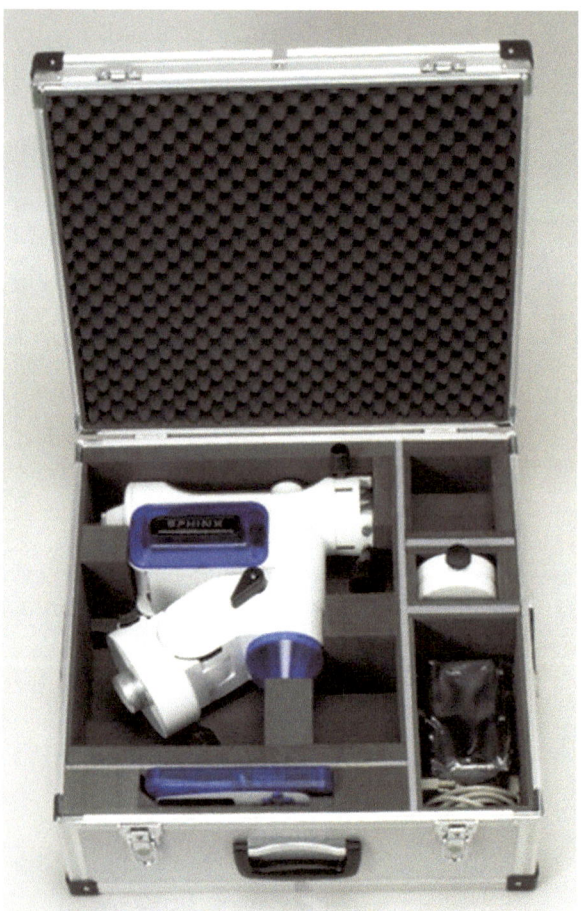

Fig. 10.16 SX aluminum carry case (Vixen)

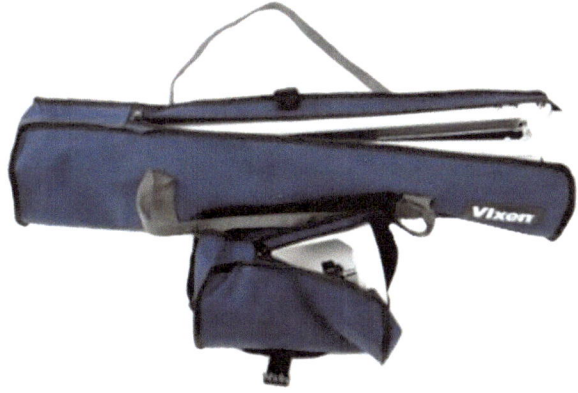

Fig. 10.17 SX tripod carrying case (Vixen)

Fig. 10.18 Star Book case (Vixen, Mr. Star Guy)

Chapter 11

Mounting Other Optical Tubes on the Vixen Mounts

All Sphinx mounts utilize the Vixen dovetail plate system for the quick and easy mounting of telescopes onto the equatorial mount. The system was introduced on the Vixen Great Polaris (GP) equatorial mounts in 1992. The dovetail system simplified the attachment and removal of telescope optical tubes from mounts, while assuring a firm and safe attachment to the mount when operational. The Vixen design has evolved into the most popular of the two standard mounting systems in the amateur astronomy world, with its only rival being the Losmondy Mounting Plate system. Many third-party suppliers now offer Vixen-compatible dovetail brackets and mounting blocks.

The basis for the Vixen mounting system is the use of a basic dovetail, as shown in Fig. 11.1a, b. Note the notch at the mid-point of the plate for securing the dovetail to the mounting block with the mounting bolt.

© Springer International Publishing Switzerland 2016
J.L. Chen, A. Chen, *The Vixen Star Book User Guide*, The Patrick Moore
Practical Astronomy Series, DOI 10.1007/978-3-319-21593-8_11

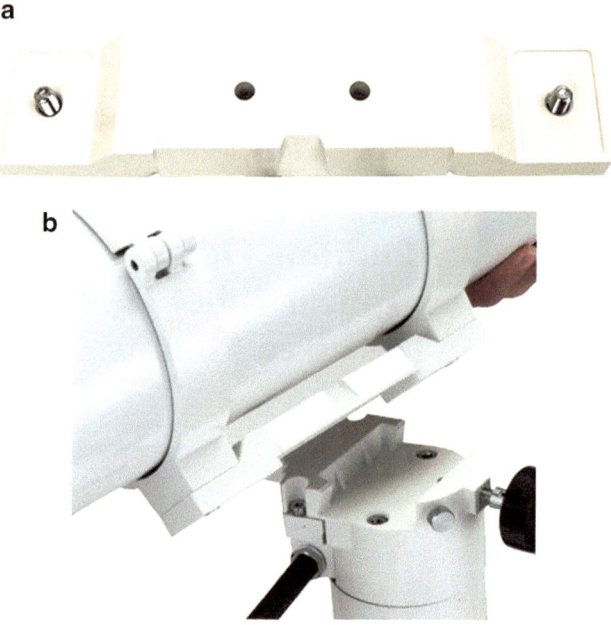

Fig. 11.1 (**a**) Basic Vixen dovetail tube plate (Vixen). (**b**) Basic Vixen dovetail with telescope mounted ready for attachment with dovetail block (Vixen)

Alternatively, Vixen offers a dovetail plate that is adaptable to various tube ring or clamshell screw hole configurations of other manufacturers (Fig. 11.2).

Fig. 11.2 Vixen universal dovetail plate (Vixen)

Other companies, such as Celestron and Stellarvue for example, also provide their telescopes with dovetail plates or rails that attach to their telescopes and fit the standard Vixen Dovetail plate mounting block.

The Mounting Block design used by Vixen features two bolts to secure the dovetail to the mount (Fig. 11.3). The central large bolt is designed to fit within a notch in the basic Vixen dovetail plate, with a small secondary angled bolt to provide stability. Both bolts work together to securely attach the telescope to the mount and prevents slippage of the dovetail that could cause equipment disaster. Many third-party brands of dovetails do not feature the notch feature.

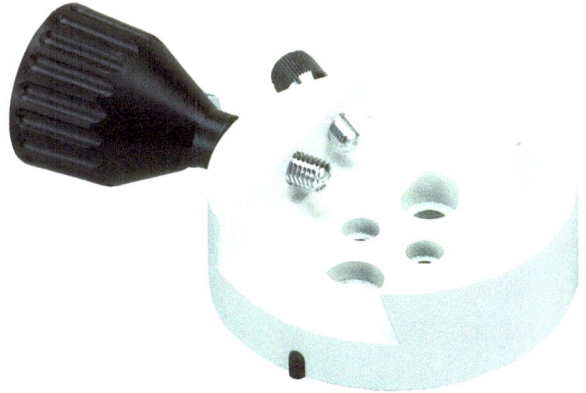

Fig. 11.3 Dovetail-plate mounting block (Vixen)

The Vixen accessory plate allows the user to mount two telescopes, a camera and telescope, or a combination of equipment on the Sphinx mount at the same time (Fig. 11.4).

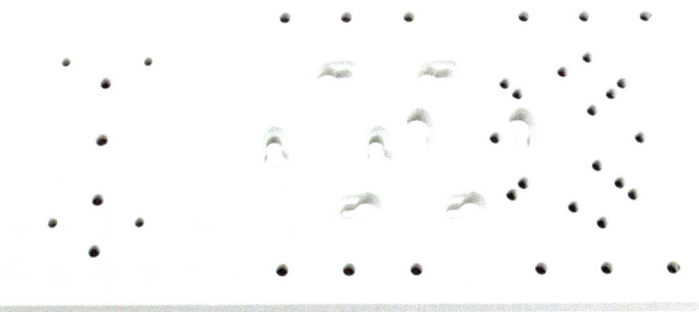

Fig. 11.4 Vixen accessory plate (Vixen)

Figure 11.5 provides a sample view of the attachment of telescope rings to the dovetail plate assembly, although, typically, the rings are attached at opposite ends of the dovetail plate.

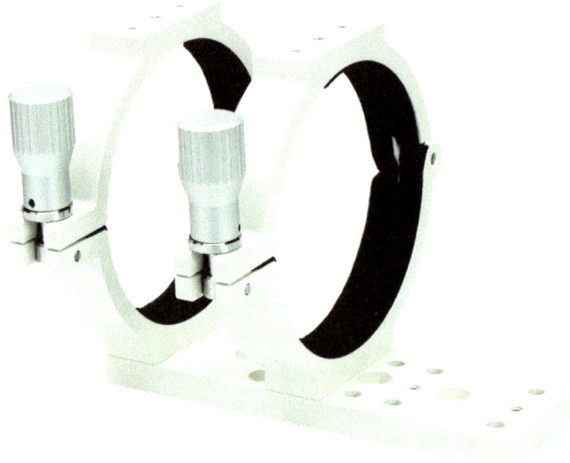

Fig. 11.5 Telescope rings on mounting plate (Vixen)

Chapter 12

Maintenance and Care of the Star Book Electronics and Sphinx Mounts

Owners of telescopes, telescope mounts, eyepieces, cameras (both digital and film), and numerous accessories have often invested enormous amounts of money in their pursuit of the science of astronomy.

Here is some guidance in the care and feeding of the Vixen Sphinx mounts and the associated Star Book computer/controllers.

It is important that owners of the Sphinx mounts and Star Book TEN, Star Book One, or original Star Book recognize that this equipment is made of commercial grade mechanical and electronic components. This is not to criticize Vixen, as the majority of commercially available alt-az and equatorial mounts, and axis drives or GoTo electronics on the market are just that: commercial grade. Components that are MIL-spec or NASA-spec would raise the price of our favorite equipment significantly to the point of not being affordable to the amateur astronomer. Nor do the needs of the average backyard astronomer require MIL-spec or NASA-spec level components.

General Maintenance and Care

The best advice is to treat the Sphinx mount and the Star Book electronics like you would treat yourself:

- Store the Sphinx and Star Book indoors. An HVAC controlled environment will protect the Sphinx and the Star Book from damage due to moisture and corrosion. The occasional star party use is okay, so long as the equipment is covered and protected from the weather when not in use.

© Springer International Publishing Switzerland 2016
J.L. Chen, A. Chen, *The Vixen Star Book User Guide*, The Patrick Moore
Practical Astronomy Series, DOI 10.1007/978-3-319-21593-8_12

- The appropriateness of a garage or an unheated/non-A/C observatory is debatable. The near-ambient temperature of an unheated garage or observatory means there is no observing time wasted waiting for the telescope optics to adjust to the outdoor ambient temperature. The concern is the presence of moisture and humidity. Corrosion of mechanical and electronic parts can lead to a shortening of the life span of astronomical equipment. A garage environment can be a somewhat better location than an unheated and uninsulated shed. Here again, the reader is reminded that the Star Book electronics and the Sphinx mounts are high quality commercial grade products. They are not MIL-spec, and are not built to operate in extreme environments.

- Beware of the environmental conditions. The official Vixen operating temperature specification for the Star Book is −20 to +40 °C, or −4 to +104 °F. At the lower temperatures, the Star Book will become non-responsiveness to commands. At extreme low temperatures, pressing the keys will yield no response. The lubricant in the Sphinx mount will thicken at extreme low temperatures, stressing the drive motors on the Sphinx mount. It is recommended that the Sphinx mount and the Star Book not be used in single digit Fahrenheit temperatures. Extreme high temperatures should also be avoided. The lubricants of the mount will thin out in extreme high triple digit Fahrenheit temperatures, causing wear on bearings and shortening the mechanical components lifetime. Extreme high triple digit Fahrenheit temperatures have an adverse effect on electronic components, with overheating causing controller failure. A trip to the repair shop should be avoided at all costs.

- Be mindful of the load placed on the Sphinx mount. A telescope that is too large for the mount will exhibit tremors and vibrations that will affect observing, while mechanically stressing the bearings and motor drives. An overloaded mount can shorten the mechanical and electrical life expectancy.

- Keep the mount in a clean, dust-free environment. Dirt and dust in mechanical components causes mechanical wear, and a layer of dust on electronic components can cause inadequate cooling, resulting in heat failure of semi-conductors and related components.

- If a problem arises, consult the Vixen dealer where the mount was purchased. Many problems are covered under warranty. If not covered under warranty, at least the proper repair parts are available.

- The discussion of warranty coverage leads to a discussion of where to purchase the Sphinx/Star Book TEN system. If it is geographically convenient, it is highly recommended that the Sphinx mount with Star Book TEN be purchased in person at an authorized Vixen dealer. Too many people try to save money by buying on-line. This is a false economy. Believe it or not, the astronomy industry is not a big money, high profit business. With the exception of two dominant major companies, many telescope businesses, either manufacturers or stores, are Mom and Pop operations run by people who love science and astronomy. They have expertise in amateur astronomy, provide quality products, provide personalized service, and are able to perform many repairs in their own shops. The smaller telescope shops struggle to compete with high volume Internet or mail-order firms who offer little or no service and rely on manufacturers to repair faulty equipment. Consumers need to understand the retail business. There are three criteria for retail competition: Quality, Service and Price. The consumer can only get two of the three. A lower price means the consumer sacrifices either service or quality. A high quality complex piece of equipment such as the Sphinx mount and the Star Book TEN controller may need the service and expertise that only an in-store expertise and in-person service can provide.

Mechanical Adjustments

There is an extensive discussion on Peter Enzerink's website, http://enzerink.net/peter/blog/2008/06/24/vixen-star-book-faq, covering the frequently asked questions on the first generation Sphinx mount, the SXW. This website contains detailed instructions of adjusting the tension of the R/A and declination axises and adjusting the worm gear of the SXW mount, much of which can be applied to the new generation of Sphinx mounts, the SX2, SXD2, and the SXP. The reader is cautioned that much of the adjustments will void the warranty. The reader who attempts these adjustments does so at his/her own risk. If the reader is not mechanically inclined and is without proper tools, performing these mechanical adjustments can cause more harm than good to their equipment. If there are problems with the Sphinx mount or the Star Book TEN, please consult with the local Vixen dealer or Vixen directly before attempting to perform any adjustments on your own. Some adjustments are best left to the experts.

Firmware Updates

Two appendices of this book are devoted to the upgrading of the firmware for the Star Book TEN and the original Star Book. The procedures are from the Vixen website. Again, this is not for the faint-of-heart. These firmware updates can only be accomplished using a PC computer. If the reader is not comfortable with performing the upgrade procedure, or does not possess computer skill or savvy, seek out the local Vixen dealer to perform the firmware upgrade. In many cases, the old adage "If it ain't broke, don't fix it" applies. If your Star Book TEN is performing to your satisfaction, the upgrade is probably not needed and does not need to be applied. Check with the version description of the firmware modification to see if it is needed.

Transporting Advice

Vixen and several other manufacturers produce carrying cases for the Sphinx mounts, tripods, and Star Book TEN and original Star Book. It is always a good idea to protect your investment. A few dollars spent on protective cases for the mount and electronics will pay dividends when the equipment arrives at a remote site or star party intact and is in good working order (Figs. 12.1, 12.2, and 12.3).

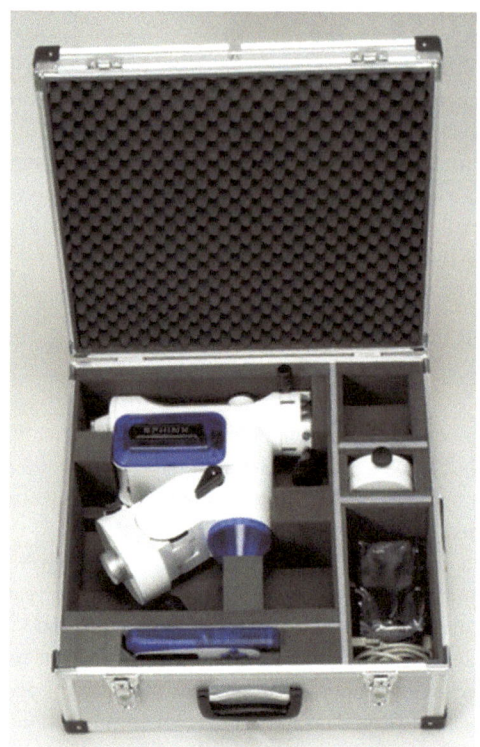

Fig. 12.1 Sphinx aluminum case (Vixen)

Fig. 12.2 Star Book case (Mr. Star Guy)

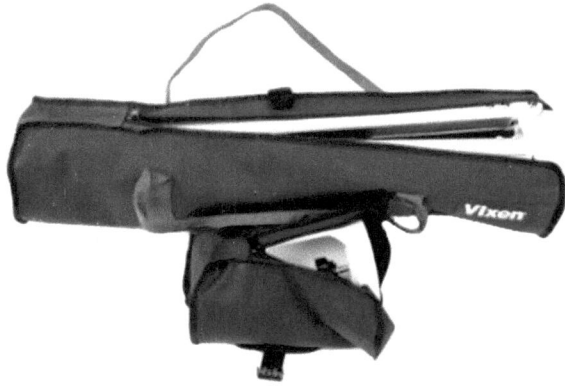

Fig. 12.3 Tripod carrying case (Mr. Star Guy)

Appendix A

Troubleshooting Checklist

From the Mr. Star Guy website:

If you are having difficulty with your Star Book system, please review these steps before calling for technical support.

Did you install the battery into the back of the Star Book? If Yes then proceed, if No, please install battery before moving forward.

Latitude: _____ Be sure it is set to "West" as it comes from the factory set for Japan Longitude: _____

Time: _____

Hours from GMT: _____ Be sure you see the—Negative symbol to the left of the Hours (for users in The Americas)

Date: _____ When Polar Aligning, start with the tripod Level

When Locking the Declination and Right Ascension Axis, be sure it is securely tightened. You may want to try this with the mount off first to see how much pressure is required to lock the axis.

Low voltage: Whenever the mount experiences low voltage you may have problems. We strongly encourage all users to use a 17 A h or more DC source OR the AC adapter to operate the mount.

© Springer International Publishing Switzerland 2016   147
J.L. Chen, A. Chen, *The Vixen Star Book User Guide*, The Patrick Moore
Practical Astronomy Series, DOI 10.1007/978-3-319-21593-8

Here is a table to help with approximate distance estimates;

Distance spanned by the little finger	1°
Distance spanned by three fingers held together	5°
Distance spanned by the closed fist	10°
Distance spanned between the little finger in pointing finger when spread apart	15°

Appendix B

Messier Catalog

During his lifetime, Charles Messier (1730–1817) was an astronomer noted for his comet discoveries. He found 13 comets and succeeded in 7 more independent co-discoveries. Messier compiled a list of deep sky objects that would were easily confused for comets to help him in his comet searches. Ironically, he is more famous today for his list of non-comet deep sky objects than his comet discoveries. Known as the Messier catalog, it contains 110 objects (actually 109 because of a duplication), including nebulae, clusters and galaxies. The list is fully contained within both the Star Book TEN and the original Star Book databases. Many of the Messier objects within the Star Book TEN database include photographs of the objects.

Messier number	Common name	Constellation	R.A. H:M.S	DEC	App Mag	Type
M1	Crab Nebula	Taurus	5:34.5	22°01′	8.4	Planetary Nebula
M2		Aquarius	21:33.5	−00°49′	6.5	Globular Cluster
M3		Canes Venatici	13:42.2	28°23′	6.4	Globular Cluster
M4		Scorpius	16:23.6	−26°32′	5.9	Globular Cluster
M5		Serpens	15:18.5	2°05′	5.8	Globular Cluster
M6	Butterfly Cluster	Scorpius	17:40.0	−32°13′	4.2	Open Cluster
M7	Ptolemy Cluster	Scorpius	17:54.0	−34°49′	3.3	Open Cluster
M8	Lagoon Nebula	Sagittarius	18:03.7	−24°23′	5.8	Emission Nebula
M9		Ophiuchus	17:19.2	−18°31′	7.9	Globular Cluster
M10		Ophiuchus	16:57.2	−4°06′	6.6	Globular Cluster
M11	Wild Duck Cluster	Scutum	18:51.1	−6°16′	5.8	Open Cluster
M12		Ophiuchus	16:47.2	−1°57′	6.6	Globular Cluster
M13	Hercules Cluster	Hercules	16:41.7	36°28′	5.9	Globular Cluster

(continued)

© Springer International Publishing Switzerland 2016
J.L. Chen, A. Chen, *The Vixen Star Book User Guide*, The Patrick Moore Practical Astronomy Series, DOI 10.1007/978-3-319-21593-8

(continued)

Messier number	Common name	Constellation	R.A. H:M.S	DEC	App Mag	Type
M14		Ophiuchus	17:37.6	−3°15′	7.6	Globular Cluster
M15		Pegasus	21:30.0	12°10′	6.4	Globular Cluster
M16	Eagle Nebula	Serpens	18:18.9	−13°47′	6	Emission Nebula
M17	Omega, Swan, Horseshoe, or Lobster Nebula	Sagittarius	18:20.8	−16°11′	7	Nebula
M18		Sagittarius	18:19.9	−17°08′	6.9	Open Cluster
M19		Ophiuchus	17:02.6	−26°16′	7.2	Globular Cluster
M20	Trifid Nebula	Sagittarius	18:02.4	−23°02′	8.5	Diffuse Nebula
M21		Sagittarius	18:04.7	−22°30′	5.9	Open Cluster
M22	Sagittarius Cluster	Sagittarius	18:36.4	−23°54′	5.1	Globular Cluster
M23		Sagittarius	17:56.9	−19°01′	5.5	Open Cluster
M24	Sagittarius Star Cloud	Sagittarius	18:16.4	−18°29′	4.5	Open Cluster
M25		Sagittarius	18:31.7	−19°15′	4.6	Open Cluster
M26		Scutum	18:45.2	−9°24′	8	Open Cluster
M27	Dumbbell Nebula	Vulpecula	19:59.6	22°43′	8.1	Planetary Nebula
M28		Sagittarius	18:24.6	−24°52′	6.9	Globular Cluster
M29		Cygnus	20:23.0	38°32′	6.6	Open Cluster
M30		Capricornus	21:40.4	−23°11′	7.5	Globular Cluster
M31	Andromeda Galaxy	Andromeda	0:42.7	41°16′	3.4	Spiral Galaxy
M32		Andromeda	0:42.7	40°52′	8.2	Elliptical Galaxy
M33	Pinwheel Galaxy	Triangulum	1:33.8	30°39′	5.7	Spiral Galaxy
M34		Perseus	2:42.0	42°47′	5.2	Open Cluster
M35		Gemini	6:08.8	24°20′	5.1	Open Cluster
M36		Auriga	5:36.3	34°08′	6	Open Cluster
M37		Auriga	5:52.0	32°33′	5.6	Open Cluster
M38		Auriga	5:28.7	35°50′	6.4	Open Cluster
M39		Cygnus	21:32.3	48°26′	4.6	Open Cluster
M40	Winnecke 4	Ursa Major	12:22.2	68°05′	8	Dbl star
M41		Canis Major	6:47.0	−20°44′	4.5	Open Cluster
M42	Great Orion Nebula	Orion	5:35.3	−5°27′	4	Nebula
M43	De Mairan's Nebula	Orion	5:35.5	−5°16′	9	Nebula
M44	Beehive Cluster	Cancer	8:40.0	19°59′	3.1	Open Cluster
M45	Pleiades	Taurus	3:47.5	24°07′	1.2	Open Cluster
M46		Puppis	7:41.8	−14°49′	6.1	Open Cluster
M47		Puppis	7:36.6	−14°30′	4.4	Open Cluster
M48		Hydra	8:13.8	−5°48′	5.8	Open Cluster
M49		Virgo	12:29.8	8°00′	8.4	Elliptical galaxy
M50		Monoceros	7:03.0	−8°20′	5.9	Open Cluster
M51	Whirlpool Galaxy	Canes Venatici	13:29.9	47°12′	8.1	Spiral Galaxy
M52		Cassiopeia	23:24.2	61°35′	6.9	Open Cluster
M53		Coma Berenices	13:12.9	18°10′	7.7	Globular Cluster
M54		Sagittarius	18:55.1	−30°29′	7.7	Globular Cluster
M55		Sagittarius	19:40 .0	−30°68′	7	Globular Cluster
M56		Lyra	19:16.6	30°11′	8.2	Globular Cluster

(continued)

(continued)

Messier number	Common name	Constellation	R.A. H:M.S	DEC	App Mag	Type
M57	Ring Nebula	Lyra	18:53.6	33°02′	9	Planetary Nebula
M68		Virgo	12:37.7	11°49′	9.8	Spiral Galaxy
M59		Virgo	12:42.0	11°39′	9.8	Elliptical Galaxy
M70		Virgo	12:43.7	11°33′	8.8	Elliptical Galaxy
M61		Virgo	12:21.9	4°28′	9.7	Spiral Galaxy
M62		Ophiuchus	17:01.2	−30°07′	6.6	Globular Cluster
M68	Sunflower Galaxy	Canes Venatici	13:15.8	42°02′	8.6	Spiral Galaxy
M64	Black Eye Galaxy	Coma Berenices	12:56.7	21°41′	8.5	Spiral Galaxy
M50	Leo's Triplet	Leo	11:18.9	13°05′	9.3	Spiral Galaxy
M66	Leo's Triplet	Leo	11:20.3	12°59′	9	Spiral Galaxy
M67		Cancer	8:50.3	11°49′	6.9	Open Cluster
M68		Hydra	12:39.5	−26°45′	8.2	Globular Cluster
M69		Sagittarius	18:31.4	−32°21′	7.7	Globular Cluster
M70		Sagittarius	18:43.2	−32°18′	8.1	Globular Cluster
M70		Sagitta	19:53.7	18°47′	8.3	Globular Cluster
M72		Aquarius	20:53.5	−12°32′	9.4	Globular Cluster
M73		Aquarius	20:68.0	−12°38′		Asterism
M74		Pisces	1:36.7	15°47′	9.2	Spiral Galaxy
M75		Sagittarius	20:06.1	−21°55′	8.6	Globular Cluster
M76	Cork Nebula, Little Dumbbell	Perseus	1:42.2	51°34′	11.5	Planetary Nebula
M77		Cetus	2:42.7	0°01′	8.8	Spiral Galaxy
M78		Orion	5:46.7	0°03′	8	Nebula
M79		Lepus	5:24.2	−24°33′	8	Globular Cluster
M80		Scorpius	16:17.0	−22°59′	7.2	Globular Cluster
M81	Bodes Nebula	Ursa Major	9:55.8	69°04′	6.8	Spiral Galaxy
M82	Cigar Galaxy	Ursa Major	9:56.2	69°41′	8.4	Irregular Galaxy.
M83	Southern Pinwheel Galaxy	Hydra	13:37.7	−29°52′	7.6	Spiral Galaxy
M84		Virgo	12:25.1	12°53′	9.3	Elliptical Galaxy
M85		Coma Berenices	12:25.4	18°11′	9.2	Elliptical Galaxy
M86		Virgo	12:26.2	12°57′	9.2	Elliptical Galaxy
M87	Virgo A	Virgo	12:30.8	12°24′	8.6	Elliptical Galaxy
M88		Coma Berenices	12:32.0	14°25′	9.5	Spiral Galaxy
M89		Virgo	12:35.7	12°33′	9.8	Elliptical Galaxy
M87		Virgo	12:36.8	13°10′	9.5	Spiral Galaxy
M91		Coma Berenices	12:35.4	14°30′	10.2	Spiral Galaxy
M92		Hercules	17:17.1	43°08′	6.5	Globular Cluster
M93		Puppis	7:44.6	−23°52′	6.2	Open Cluster
M94		Canes Venatici	12:50.9	41°07′	8.1	Spiral Galaxy
M95		Leo	10:44.0	11°42′	9.7	Spiral Galaxy
M96		Leo	10:46.8	11°49′	9.2	Spiral Galaxy
M97	Owl Nebula	Ursa Major	11:14.9	55°01′	11	Planetary Nebula
M98		Coma Berenices	12:13.8	14°54′	10.1	Spiral Galaxy
M87	Pin Wheel Nebula	Coma Berenices	12:18.8	14°25′	9.8	Spiral Galaxy

(continued)

(continued)

Messier number	Common name	Constellation	R.A. H:M.S	DEC	App Mag	Type
M100		Coma Berenices	12:22.9	15°49′	9.4	Spiral Galaxy
M101		Ursa Major	14:03.2	54°21′	7.7	Spiral Galaxy
M102	Probably M101 duplicate		14:03.2	54°21′	7.7	Duplicate
M103		Cassiopeia	1:33.1	70°42′	7.4	Open Cluster
M104	Sombrero Galaxy	Virgo	12:40.0	−11°37′	8.3	Spiral Galaxy
M105		Leo	10:47.9	12°35′	9.3	Elliptical Galaxy
M106		Canes Venatici	12:19.0	47°18′	8.3	Spiral Galaxy
M107		Ophiuchus	16:32.5	−13°03′	8.1	Globular Cluster
M108		Ursa Major	11:11.6	55°40′	10	Spiral Galaxy
M109		Ursa Major	11:57.7	53°23′	9.8	Spiral Galaxy
M110		Andromeda	0:40.3	41°41′	8	Elliptical Galaxy

Appendix C

Selected Non-Messier Catalog NGC Objects

Both the Star Book TEN and the original Star Book contain J.L.E. Dreyer's New General Catalogue of Nebulas and Cluster of Stars (NGC) and the two supplements, the Index Catalogues (IC) in their databases.

The NGC list, compiled in 1888, contains 7840 objects. The two supplements were published in 1895 and 1908 respectively, with the first containing 1520 additional deep sky objects and the second containing 3866 additional IC objects.

Many of the objects listed in the NGC/IC database of the both Star Books are not visible visually through the size of telescope that can be accommodated by the Sphinx family of mounts. It is possible to detect and image the fainter objects photographically.

Unlike the much shorter Messier Catalog of Appendix B, it is impractical to list the thousands of NGC/IC objects in this book.

However, here are some NGC and IC objects that are recommended. This list is the called the SAA 100.

A discussion thread began in 2000 in the sci.astro.amateur (SAA) newsgroup when a question was posted: "What are your favorite non-Messier objects for 8–12″ telescopes?" The SAA Newsgroup participants responded enthusiastically to the question, posting many messages nominating a wide variety of objects.

The table below lists the SAA 100 in rank order by number of votes received. About half the objects received only one vote each; these are listed alphabetically at the end of the list.

This is by no means the only available list of non-Messier objects, with the Caldwell List coming quickly to mind. The SAA 100 is a good starting point for readers of this book, with many of these objects accessible to telescopes mounted on the Sphinx/Star Book mounts.

© Springer International Publishing Switzerland 2016   153
J.L. Chen, A. Chen, *The Vixen Star Book User Guide*, The Patrick Moore
Practical Astronomy Series, DOI 10.1007/978-3-319-21593-8

SAA 100

Best Non-Messier Objects, in Rank Order

Object	Type	Con	Visual Mag	Size	RA	Dec	Pop. name	Notes
NGC 253	Gal	Scl	7.2	25.0′×7.0′	00 h 47 m 35 s	−25° 17′ 01″		Edge-on spiral
NGC 4565	Gal	Com	9.6	15.5′×1.9′	12 h 36 m 20s	+25° 59′ 23″	Bernice's Hair Clip	Classic edge-on spiral with dust lane
NGC 6960	SNR	Cyg		70.0′×6.0′	20 h 45 m 38 s	+30° 43′ 20″	Western Veil	
NGC 6992	SNR	Cyg		25.0′×20.0′	20 h 56 m 14 s	+31° 04′ 20″	Eastern Veil	
NGC 869	OC	Per	5.3	30.0′	02 h 19 m 03 s	+57° 08′ 58″	Double Cluster	w/NGC 884
NGC 884	OC	Per	6.1	30.0′	02 h 22 m 27 s	+57° 06′ 57″	Double Cluster	w/NGC 869
NGC 457	OC	Cas	6.4	13.0′	01 h 19 m 10s	+58° 20′ 02″	Owl Cluster	
NGC 5139	GC	Cen	3.7	36.3′	13 h 26 m 46 s	−47° 28′ 45″	Omega Centauri	Best GC in the sky
NGC 7293	PN	Aqr	6.3	16.0′×12.0′	22 h 29 m 40 s	−20° 47′ 23″	Helical Nebula	
NGC 7789	OC	Cas	6.7	16.0′	23 h 57 m 04 s	+56° 44′ 09″		
NGC 2237	BN	Mon	5.5	70.0′×80.0′	06 h 32 m 19 s	+04° 59′ 03″	Rosette Nebula	OC NGC 2244 embedded in nebula
NGC 2244	OC	Mon	4.8	24.0′	06 h 32 m 25 s	+04° 52′ 03″		Involved with Rosette Neb. (NGC 2237)
NGC 2359	BN	CMa		8.0′	07 h 17 m 48 s	−13° 12′ 54″	Thor's Helmet; Duck Nebula	Wolf-Rayet remnant
NGC 2392	PN	Gem	8.6	47.0″×43.0″	07 h 29 m 10 s	+20° 54′ 42″	Eskimo Nebula; Clown Face	
NGC 3242	PN	Hya	8.6	40.0″×35.0″	10 h 24 m 48 s	−18° 38′ 14″	Ghost of Jupiter	
NGC 6543	PN	Dra	8.3	22.0″×16.0″	17 h 58 m 36 s	+66° 38′ 17″	Cat's Eye Nebula	
NGC 4631	Gal	CVn	9.2	17.0′×3.5′	12 h 42 m 11 s	+32° 32′ 42″		Same LP field as NGC 4656
NGC 4656	Gal	CVn	10.5	22.0′×3.0′	12 h 43 m 58 s	+32° 10′ 21″		same LP field as NGC 4631
NGC 5128	Gal	Cen	6.8	18.2′×14.5′	13 h 25 m 29 s	−43° 01′ 07″	Centaurus A	Strong radio source
NGC 6781	PN	Aql	11.8	1.9′×1.8′	19 h 18 m 28 s	+06° 32′ 46″		
NGC 6826	PN	Cyg	8.8	27.0″×24.0″	19 h 44 m 53 s	+50° 31′ 42″	Blinking Planetary	
NGC 7009	PN	Aqr	8.3	28.0″×23.0″	21 h 04 m 15 s	−11° 21′ 49″	Saturn Nebula	
Abell 1656	Gal cluster	Com	11.0	120.0′	12 h 59 m 48 s	+27° 59′ 04″	Coma Gal Cluster	Tough in 8–12″ aperture
NGC 1023	Gal	Per	9.4	9.0′×4.0′	02 h 40 m 27 s	+39° 03′ 47″		
NGC 2362	OC	CMa	4.1	8.0′	07 h 18 m 48 s	−24° 56′ 51″		
NGC 2403	Gal	Cam	8.5	17.8′	07 h 36 m 55 s	+65° 35′ 42″		
NGC 4038	Gal	Cor	10.3	2.6′×1.8′	12 h 01 m 53 s	−18° 51′ 55″	The Antennae; Ringtail Gal	Interacting with NGC 4039
NGC 4039	Gal	Cor	10.6	3.2′×2.2′	12 h 01 m 54 s	−18° 53′ 07″	The Antennae	Interacting with NGC 4038

(continued)

(continued)

Object	Type	Con	Visual Mag	Size	RA	Dec	Pop. name	Notes
NGC 5907	Gal	Dra	10.3	12.8′×1.8′	15 h 15 m 52 s	+56° 19′ 48″		
NGC 6369	PN	Oph	11.0	30.0″×29.0″	17 h 29 m 22 s	−23° 45′ 37″		
NGC 663	OC	Cas	7.1	16.0′	01 h 46 m 04 s	+61° 15′ 00″		
NGC 654	OC	Cas	6.5	5.0′	01 h 44 m 10 s	+61° 53′ 00″		
NGC 659	OC	Cas	7.9	5.0′	01 h 44 m 16 s	+60° 42′ 00″		
NGC 7000	BN	Cyg		175.0′×110.0′	20 h 58 m 32 s	+44° 33′ 21″	North American Nebula	Large; often easier in binoculars than telescope
NGC 7331	Gal	Peg	9.5	11.4′×4.0′	22 h 37 m 08 s	+34° 25′ 27″	Little And Gal	
NGC 7662	PN	And	8.6	17.0″×14.0″	23 h 25 m 57 s	+42° 32′ 44″	Blue Snowball Nebula	
B 59, 65-7	DN	Oph		300.0′	17 h 21 m 02 s	−26° 59′ 58″	Pipe Nebula (stem)	
B 78	DN	Oph		200.0′	17 h 33 m 02 s	−25° 59′ 58″	Pipe Nebula (bowl)	
IC 1396	BN	Cep	3.5	154.0′×140.0′	21 h 39 m 09 s	+57° 46′ 58″		
IC 418	PN	Lep	10.7	14.0″×11.0″	05 h 27 m 30 s	−12° 41′ 32″		
IC 4665	OC	Oph	4.2	41.0′	17 h 46 m 20s	+05° 43′ 08″		
Mel 111	OC	Com	1.8	275.0′	12 h 25 m 00 s	+26° 00′ 07″	Coma Berenices Star Cluster	
Mel 20	OC	Per	1.2	185.0′	03 h 22 m 03 s	+48° 59′ 56″	Alpha Persei Association	
NGC 1502	OC	Cam	6.9	8.0′	04 h 07 m 45 s	+62° 19′ 49″		Near SE end of Kemble's Cascade
NGC 1528	OC	Per	6.4	24.0′	04 h 15 m 24 s	+51° 13′ 49″		
NGC 1907	OC	Aur	8.2	7.0′	05 h 28 m 00 s	+35° 18′ 53″		
NGC 1973	BN	Ori		5.0′×5.0′	05 h 35 m 09 s	−04° 43′ 56″	Part of Running Man Nebula	
NGC 1975	BN	Ori		10.0′×5.0′	05 h 35 m 21 s	−04° 40′ 56″	Part of Running Man Nebula	
NGC 1977	BN	Ori		20.0′×10.0′	05 h 35 m 27 s	−04° 49′ 56″	Part of Running Man Nebula	42 Orionis nebula
NGC 2070	BN	Dor	8.3	5.0′	05 h 38 m 39 s	−69° 04′ 51″	Tarantula Nebula	In Lg. Magellanic Cloud
3C 273	Quasar	Vir	12.0		12 h 29 m 06 s	+02° 03′ 01″		Brightest quasar; most remote object visible in modest amateur telescopes (~2 billion light years)
Albireo	Star	Cyg	3.1		19 h 30 m 45 s	+27° 57′ 55″		Superb double star; blue-white/yellow

(continued)

(continued)

Object	Type	Con	Visual Mag	Size	RA	Dec	Pop. name	Notes
Cr 399	Asterism	Vul	3.6	60.0′	19 h 25 m 26 s	+20° 11′ 18″	Brocchi's Cluster; The Coathanger	Once assumed to be a OC; data from Hipparcos spacecraft shows it to be a chance alignment of stars
Fornax Gal. Cluster	Gal cluster	For		3°×2°	03 h 38 m 31 s	−35° 26′ 40″		Approx. 20 galaxies brighter than mag. 13
Kemble's Cascade	Asterism	Cam			03 h 57 m 30 s	+63° 04′ 13″		First described by Canadian amateur Lucian J. Kemble; beautiful chain of about 20 mag. 5…9 stars; coordinates are for SAO 12969, a mag. 5 star in the middle of the Cascade
King 10	OC	Cep		3.0′	22 h 54 m 58 s	+59° 10′ 16″		
Markarian's Chain	Gal chain	Vir			12 h 25 m 04 s	+12° 53′ 16″		String of bright galaxies; covers 3° of sky, starting with M84 and M86 in Virgo, ending with NGCs 4459 and 4474 in Coma Berenices; coordinates are for M 84
Mel 25	OC	Tau	0.5	330.0′	04 h 27 m 02 s	+16° 00′ 03″	Hyades	Aldebaran not a member
NGC 104	GC	Tuc	4.0	30.9′	00 h 24 m 10 s	−72° 04′ 37″	47 Tucanae	
NGC 1535	PN	Eri	10.4	20.0″×17.0″	04 h 14 m 16 s	−12° 44′ 16″		Multiple shells
NGC 2158	OC	Gem	8.6	5.0′	06 h 07 m 33 s	+24° 05′ 56″		
NGC 2169	OC	Ori	5.9	7.0′	06 h 08 m 27 s	+13° 56′ 59″	"37" Cluster	
NGC 2174	BN	Ori		25.0′×20.0′	06 h 10 m 01 s	+20° 33′ 58″		
NGC 2232	OC	Mon	3.9	30.0′	06 h 26 m 37 s	−04° 44′ 54″		
NGC 225	OC	Cas	7.0	12.0′	00 h 43 m 28 s	+61° 47′ 06″		
NGC 2261	BN	Mon		2.0′×1.0′	06 h 39 m 13 s	+08° 44′ 01″	Hubble's Variable Nebula	
NGC 2264	OC	Mon	3.9	30.0′×60.0′	06 h 40 m 58 s	+09° 53′ 42″	Christmas Tree Cluster; Cone Nebula	Includes naked-eye S Mon (15 Mon)
NGC 2301	OC	Mon	6.0	12.0′	06 h 51 m 49 s	+00° 28′ 04″		
NGC 2360	OC	CMa	7.2	13.0′	07 h 17 m 48 s	−15° 36′ 53″		
NGC 2438	PN	Pup	11.0	1.1′	07 h 41 m 51 s	−14° 44′ 06″		In foreground of M 46

(continued)

(continued)

Object	Type	Con	Visual Mag	Size	RA	Dec	Pop. name	Notes
NGC 2467	BN	Pup	7.1	15.0′	07 h 52 m 30 s	−26° 22′ 52″		Use UHC or O-III filter; includes loose cluster of mag. 8–12 stars
NGC 247	Gal	Cet	9.1	20.0′×7.0′	00 h 47 m 11 s	−20° 45′ 21″		
NGC 2841	Gal	Uma	9.2	7.4′×3.5′	09 h 22 m 01 s	+50° 58′ 21″		
NGC 2903	Gal	Leo	9.0	13.3′×6.0′	09 h 32 m 10 s	+21° 29′ 58″		
NGC 3115	Gal	Sex	8.9	8.3′×3.2′	10 h 05 m 14 s	−07° 43′ 06″	Spindle Gal	
NGC 3372	BN	Car		120.0′×120.0′	10 h 43 m 47 s	−59° 52′ 01″	Eta Carina Nebula	
NGC 3532	OC	Car	3.0	55.0′	11 h 06 m 23 s	−58° 40′ 03″		
NGC 3766	OC	Cen	5.3	12.0′	11 h 36 m 05 s	−61° 37′ 04″		
NGC 3877	Gal	Uma	11.0	5.6′×1.2′	11 h 46 m 07 s	+47° 29′ 37″		
NGC 40	PN	Cep	10.7	1.0′×0.7′	00 h 13 m 08 s	+72° 31′ 47″		
NGC 4244	Gal	CVn	10.4	18.5′×2.3′	12 h 17 m 29 s	+37° 48′ 28″		
NGC 4361	PN	Cor	10.3	1.3′	12 h 24 m 30 s	−18° 47′ 38″		
NGC 4526	Gal	Vir	9.7	7.0′×2.7′	12 h 34 m 03 s	+07° 42′ 03″	Lost Gal	
NGC 4567	Gal	Vir	11.3	3.0′×2.5′	12 h 36 m 33 s	+11° 15′ 33″	Siamese Twins	Overlaps NGC 4568
NGC 4568	Gal	Vir	10.8	5.1′×2.4′	12 h 36 m 35 s	+11° 14′ 17″	Siamese Twins	Overlaps NGC 4567
NGC 4755	OC	Cru	4.2	10.0′	12 h 53 m 35 s	−60° 20′ 08″	Jewel Box Cluster; Kappa Crucis	
NGC 5746	Gal	Vir	10.3	7.4′×1.1′	14 h 44 m 57 s	+01° 57′ 20″		
NGC 6210	PN	Her	9.7	20.0″×13.0″	16 h 44 m 30 s	+23° 48′ 46″		
NGC 6231	OC	Sco	2.6	15.0′	16 h 54 m 01 s	−41° 48′ 06″	Table of Scorpius	Zeta Sco complex
NGC 6397	GC	Ara	5.7	25.7′	17 h 40 m 43 s	−53° 40′ 33″		One of the nearest Globulars
NGC 6545	Gal	Pav	13.2	1.0′×0.9′	18 h 12 m 18 s	−63° 46′ 45″	Needle Galaxy	
NGC 6572	PN	Oph	9.0	15.0″×12.0″	18 h 12 m 09 s	+06° 51′ 01″		
NGC 6633	OC	Oph	4.6	27.0′	18 h 27 m 43 s	+06° 34′ 14″		
NGC 6819	OC	Cyg	7.3	5.0′	19 h 41 m 20 s	+40° 11′ 22″		
NGC 6885	OC	Vul	8.1	7.0′	20 h 12 m 02 s	+26° 29′ 20″		
NGC 6888	BN	Cyg		20.0′×10.0′	20 h 12 m 14 s	+38° 20′ 21″	Crescent Nebula	
NGC 6939	OC	Cep	7.8	8.0′	20 h 31 m 27 s	+60° 38′ 22″		
NGC 752	OC	And	5.7	50.0′	01 h 57 m 51 s	+37° 41′ 05″		
NGC 891	Gal	And	9.9	14.0′×3.0′	02 h 22 m 36 s	+42° 20′ 50″		Edge-on spiral w/prominent dust lane
Stock 2	OC	Cas	4.4	60.0′	02 h 15 m 04 s	+59° 15′ 58″	Muscleman Cluster	

Object types: *Gal* galaxy, *OC* open cluster, *GC* globular cluster, *PN* planetary nebula, *BN* bright nebula, *DN* dark nebula, *SNR* supernova remnant

Appendix D

Websites for Comet and Satellite Orbital Elements

Comet Orbital Element Websites

The International Astronomical Union—Minor Planet Center

http://www.minorplanetcenter.net/iau/Ephemerides/Comets/index.html
In-The Sky.Org
https://in-the-sky.org/comets.php?country=240&reg1=3645&reg2=33718&t
   own=16908
Amateur Observer's Program
http://aop.astro.umd.edu/charts/index.shtml

Satellite Orbital Element Websites

Visual Satellite Observer's Home Page

http://www.satobs.org
NORAD Two-Line Element Sets Current Data
http://www.celestrak.com/NORAD/elements/
Heavens Above—Satellite and ISS Orbital Elements, Iridium flares,

http://www.heavens-above.com

© Springer International Publishing Switzerland 2016
J.L. Chen, A. Chen, *The Vixen Star Book User Guide*, The Patrick Moore
Practical Astronomy Series, DOI 10.1007/978-3-319-21593-8

Appendix E

Glossary

AltAz	Altitude and Azimuth mount. Features the intuitive left-right and up-down movements of the telescope.
AFOV	Apparent Field of View. Usually applied to telescope eyepieces.
Autoguider	This function processes the signal from a CCD camera installed on a guide scope, and it automatically guides the telescope and mount with high precision over an extended period. This enables long exposure photography and imaging of astronomical objects.
Backlash Compensation	Provides a reduced time lag at the point of revised motion where the mount drive gears briefly lose contact.
Equatorial mount	Features the ability to track an astronomical object by countering the rotation of the Earth. The RA, or right ascension, axis is set parallel to the Earth's axis. The declination axis is the axis of rotation that is at right angles to the polar axis of an equatorial mounting and that permits pointing the telescope to celestial objects of different declinations. Declination is the measurement of an objects angular distance from the celestial equator.

© Springer International Publishing Switzerland 2016

J.L. Chen, A. Chen, *The Vixen Star Book User Guide*, The Patrick Moore
Practical Astronomy Series, DOI 10.1007/978-3-319-21593-8

FOV	Field-of-view. The true FOV is found by diving the AFOV of an eyepiece by the magnification that results from using the eyepiece.
GoTo Mount	Computerized telescope mounts that automatically point the telescope towards the requested object.
HVAC	Heating, Ventilation, and Air Conditioning
IC Catalog	5386 Deep sky objects cataloged by J.L.E. Dreyer as a supplement to the NGC catalog.
Kings rate	The tracking rate developed to account for atmospheric refraction.
Meridian	An imaginary line drawn from due South directly overhead to due North.
Messier catalog	A list of 110 (actually 109) deep sky objects created by Charles Messier in the late 1700s. It consists of 39 galaxies, 7 nebulae, 5 planetary nebulae, and 55 star clusters.
Moon or lunar rate	Sidereal rate plus compensation for the Moon's motion around the Earth.
NGC or New General Catalogue	A catalog of deep sky objects based on William Herschel's Catalog of Nebulae. The NGC catalog contains 7840 objects, and was created by J.L.E. Dreyer.
PEC	Periodic Error Correction, which compensates for slight manufacturing errors in the mechanical gear of equatorial mount drive systems that causes an irregular motion of the tracking gear. PEC enables smoother tracking, especially for astrophotography and astro-imaging.
Sidereal rate	The standard tracking rate for compensating for the Earth's rotation. This is the rate the stars move across the sky.
Slew, or Slewing	The movement of a telescope on its mount using its drive motors.
Solar rate	The tracking rate required to maintain the accurate tracking of the Sun. This rate differs from the sidereal rate.
Star Book	The advanced astronomical navigation controller for the Vixen Sphinx SXW and SXD equatorial mounts, featuring user-friendly star chart GoTo operation.
Star Book One	A simplified controller for Sphinx family of mounts without GoTo capability.

Star Book Type S	The monochrome LCD display version of the Star Book at reduced physical dimensions with the same functionality and firmware.
Star Book TEN	The new and evolutionary improved astronomical navigation controller with high definition screen for use on second generation Vixen Sphinx mounts SX2, SXD2, SXP, and AXD mounts. The Star Book TEN features user-friendly star chart GoTo operation and improved man-machine interface.
TEN	In Japanese means Heavens.
TOTOTA	The author's acronym for turning off, turning back on, and trying the alignment process again.
Vixen dovetail	An almost universally accepted method of attaching telescope optical tubes to telescope mounts, originally introduced by Vixen.
Wall wart	An AC-to-DC power supply that plugs into a wall electrical socket.

Appendix F

Star Book TEN Updates

As of this writing, the most recent firmware build for the Star Book TEN is Version 3.40 of Dec. 4, 2014. The firmware version of the Star Book TEN controller used for the writing of this book is version 3.30.

In most cases, it is not necessary for the user to continually update the firmware of the Star Book TEN with every iteration. The author's recommendation is if the Star Book TEN is performing to the needs of the user, leave it alone. Performing the firmware update is not for the faint of heart and the computing neophyte. It is highly recommended that the user has the update performed by the Vixen dealer or by Vixen itself.

All updates can only be performed using a PC computer. This author uses an Apple iMac running iOS 10.9.5, and as a result did not and cannot perform any updates to the Star Book or Star Book TEN firmware. The author cannot verify or validate the Vixen procedures as presented.

Firmware Updates History

Below are the updates that have occurred since the introduction of the Star Book TEN. With the exception of the first four updates on the list, all updates are included in Version 3.30:

Updates Artificial Satellite Menu
A method of calculating an orbital path of artificial satellites was reviewed to perform more accurate slewing to and tracking the artificial satellites.

Updates Location Indication
It is modified that the name of location is shown entirely from beginning to end without omission in the initial setting screen directly after turning on the power.

© Springer International Publishing Switzerland 2016   165
J.L. Chen, A. Chen, *The Vixen Star Book User Guide*, The Patrick Moore
Practical Astronomy Series, DOI 10.1007/978-3-319-21593-8

Updates Parameter of Jovian Red Spot
The longitudinal parameter of Jovian Red Spot was amended and the accuracy of indication has been improved.

Amendment of Program
The following errors are amended with the new program.

– Corrects a phenomenon that some stars are not indicated in the star chart by rare chance.
– Corrects an error of resetting PEC count which occurs rarely during PEC recording.

French language is added to Setting "言語/Language"
Now French language can be chosen from among six different languages on the STAR BOOK TEN.

Saving and Reading out Setting Values
Various settings on the STAR BOOK TEN can be saved on an SD card (sold separately). The stored data are available with the STAR BOOK TEN.

Saving and Reading out PEC Data
The recoded PEC data (PPEC data) on the STAR BOOK TEN can be saved on an SD card (sold separately). The stored data are available with the STAR BOOK TEN.
Saving and Reading out Data of Orbital Elements for comet, satellite and user defined object
The registered data of the orbital elements for comet, satellite and user defined object can be saved on an SD card (sold separately). The stored data are available with the STAR BOOK TEN.

Updating STAR BOOK TEN with use of an SD card
It allows you to update your STAR BOOK TEN using an SD card (sold separately) that contains the update program.

Saving and Reading out Data with use of a PC
It allows you to transfer various setting data, PEC data and orbital elements data between your STAR BOOK TEN and a PC with a LAN crossover cable connection.

Software improvement
The icon that indicates the level of battery discharge can be seen clearer in the night vision screen.
The motor control becomes smoother at a lower 2 or 1 setting with the motor power menu.
• An Advance Unit sold separately must be installed in your STAR BOOK TEN before you use these new functions.

STAR BOOK TEN for ASCOM
With this update, the STAR BOOK TEN will work ASCOM application software.
ASCOM is open standard software for controlling astronomical equipment and instruments. It allows for controls of mounts, imaging devices, focusers and domes

in a unified operation environment. Various free drivers are available for a wide variety of astronomical equipment and instruments.

Note:

- ASCOM Standard for Astronomy was used to develop this update and Vixen is not responsible for ASCOM.
- ASCOM application software is not available from Vixen. Obtain commercially available ASCOM application software.
- Installation of this update program does not mean that it runs ASCOM application software. Both a Star Book ASCOM driver supplied by Vixen separately and ASCOM Platform (6SP1) must be installed in your PC before you use ASCOM application software.

Compatibility with SXP equatorial mounts

The STAR BOOK TEN automatically identifies the SXP mount when connected.

Improved appearances of planets

Images of some planets are more accurately displayed.

- The appearance of the waxing and waning of Mercury and Venus is displayed.
- The apparent size of Mercury, Venus, Mars, Jupiter and Saturn changes in relation to the zoom level of the screen.
- The appearance of the inclination of Saturn's ring is displayed.
- The appearance of Jupiter's stripes is improved. The approximate location of the great red spot is displayed.
- The four Galilean moons are displayed.

Improved images of a lunar eclipse

The shadow of the earth (the area of the sun's disc blocked by the earth) at a lunar eclipse is displayed.

Numerical key entry

Every parameter in the dialog boxes which are designated by number can be entered with numerical keys.

Additional languages

Languages selection is available in Italian, German and Spanish in addition to English and Japanese.

Improvement in the initial setting procedure

In the "Initial Configuration" menu, values entered in the dialog box were not saved before you advance the screen, except for "Setting local Time". This issue has been resolved and the values are saved as you enter OK in the dialog box.

Addition of Auto IP functionality for LAN

If a DHCP server is not detected on connection to a LAN, the STAR BOOK TEN automatically assigns its own IP address 169.25.a.b (where a and b are arbitrary numbers). Your steps to an update become easier.

Adding a heading "About LAN" to "Initial Configuration" and "System Menu"

The Heading "About LAN" is added to the screens of the "Initial Configuration" and "System Menu" menus to allow for finding the IP Address easily. As a result, information on the LAN is independent of the "About StarBook TEN" screen.

Applicable to StellaNavigator Ver9 (9.0c) software

Enables operation of the AXD equatorial mount through astronomical software "StellaNavigator Ver.9 (9.0c)" published by AstroArts Inc, Japan (Japanese language only).

Adding "Objects Located Recently" to "Object Menu"

The most 20 recently viewed objects chosen from "Object Menu" are stored in the memory of the STAR BOOK TEN. The memory is alive after you turn off the power switch of the AXD equatorial mount and you can refer to them as your GoTo target of your next observing session. The stored data of the objects can be erased by initializing the memory in "System menu".

Adding "Moon Map" in the "Moon" dialog box of "Sun Moon Planet"

A moon map can be displayed on the screen. The major lunar "seas" and terrains of the surface of the moon can be selected as your target object. The map of the moon can be turned around and mirror-reversed according to your viewing orientation with a diagonal prism or a diagonal mirror.

Calling up "Moon Map" by pressing the (1 SOLAR) key a little longer

The "Moon Map" function is allocated to the (1 SOLAR) key and this can be called up directly by pressing the key a little longer.

Calling up "Display Style" by pressing the (3 R/X/A) key a little longer

The "Display Style" function is allocated to the (3 R/X/A) key and this can be called up directly by pressing the key a little longer. You can go to this function from "Chart Setting" in the System menu as well.

Calling up "Bayer Designation" by pressing the (6 STAR) key a little longer

The "Bayer Designation" function is allocated to the (6 STAR) key and this can be called up directly by pressing the key a little longer. You can go to this function from "Constellation" in the Object menu as well.

Visualizing input signals from an external auto guider by lightening relevant direction keys

The direction keys are backlit in correspondence to input from an external auto guider.

More refined performance

We have revised programs to enhance stable motion.

Update Procedure

The following is taken from the Vixen Co. Japan website (www.vixen.co.jp). The author is not responsible for its contents. The photos of the screen were also taken from the website, and may not be readable in this book. Again, the author stresses before attempting any update, please refer to the Vixen website for the most up-to-date procedures and the most clear pictures.

The following exemption is posted on the Vixen website. This author does not take any responsibility for any damage experienced by a user of the Star Book TEN following these instructions:

In case of any damage experienced by a customer when installing or using this software or any charge sent by a third party to a customer due to the damage, Vixen or their dealer do not take responsibility unless the customer is not responsible and Vixen or their dealer was negligent.

Before Starting Up

First, you need to download the update program available from Vixen's website to your PC. Connecting your STAR BOOK TEN to the PC with a LAN crossover cable by one to one connection makes you ready for the update.

Use a power supply that is for the AXD equatorial mount for the update.

Make sure that you have a stable power supply for your PC to transfer the update program.

The STAR BOOK TEN may fail to reboot if the power supply is turned off during the update. In that case, ask your local Vixen dealer for repair (It is chargeable to you). All your data stored in the STAR BOOK TEN may be lost after repair.

You may have to alter the network settings in your PC for the update. Please record the original settings in the PC before changing them for the update.

Requirements

STAR BOOK TEN controller

PC (Computer) which is capable of LAN connection

LAN crossover cable

Power supply for STAR BOOK TEN (AC Adapter 12 V 3 A or Portable Power Supply SG1000SX, or equivalent.)

System Requirements for PC

Applicable OS environment:	Microsoft Windows®XP Home Edition, XP Professional, Vista, 7 Home Edition, 7 Professional, 7 Ultimate, 8, 8pro
Web Browser:	Internet Explorer version 5.0 or higher
CPU:	Pentium®II 400 MHz or better
RAM:	256 MB or more
LAN:	10BASE-T/100BASE-T

※Depending on the OS used, necessary information may not be displayed properly. If this is the case, update your OS to the latest version using Windows Update.
※Microsoft and Windows are registered trademarks of Microsoft Corporation in the US and some Countries.
All product and company names mentioned in the instructions are trademarks or registered trademarks of the respective companies.

How to Install the Update Program

Download the latest firmware version from STAR BOOK TEN Updates to a directory on your PC (Fig. F.1).

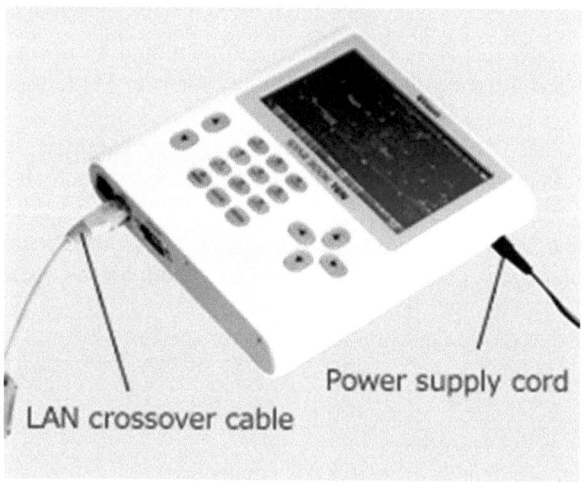

Fig. F.1 LAN and Power cable connections (Vixen)

Some data such as observing locations and defined by user that are stored in the STAR BOOK TEN may be initialized as a result of the update. You should write down your current settings before you begin to update.

Directly connect the STAR BOOK TEN to the PC with a LAN crossover cable.
※Do not power on the STAR BOOK TEN and PC at this early stage.

Power up the STAR BOOK TEN by directly plugging the 12 V power cable with center plus polarity into the unit. Turn on the PC.

Check the firmware version in your STAR BOOK TEN with the "Initial Configuration" menu that appears on the initial screen.

Procedure in Windows 8

Right-click on the background in the start up screen. Click "All apps" in the bottom right of the screen and display every application (Figs. F.2 and F.3).

Fig. F.2 Windows 8 Start screen (Vixen)

Fig. F.3 Windows 8 Start screen 2 (Vixen)

Select "Control Panel" and go to "Network and Internet" (Figs. F.4 and F.5).

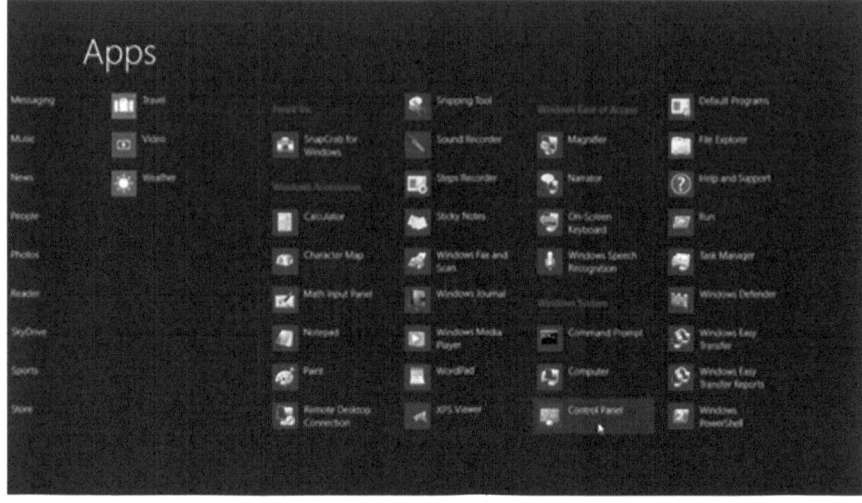

Fig. F.4 Windows 8 Apps screen (Vixen)

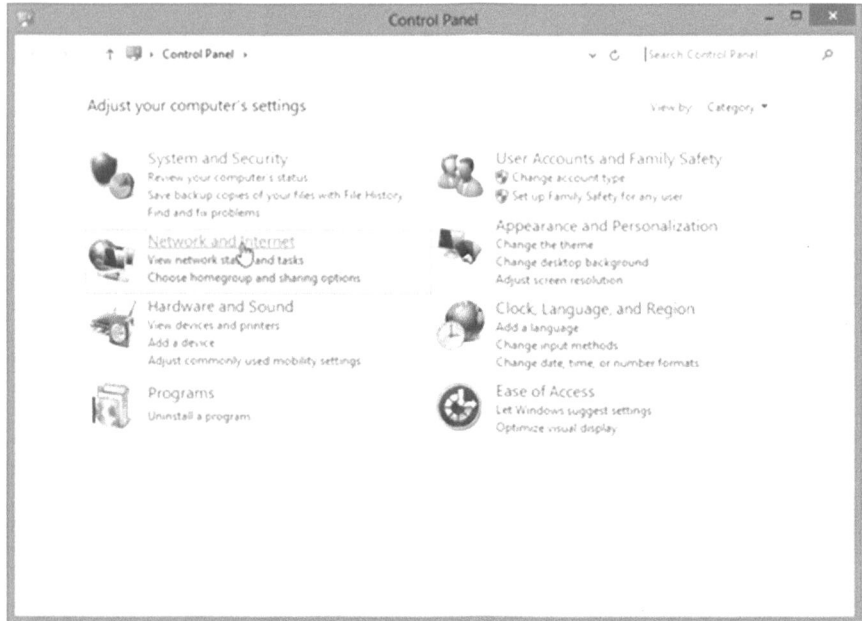

Fig. F.5 Windows 8 Network and Internet select screen (Vixen)

Click on "Network and Sharing Center" The "View your basic network information and set up connections" screen is displayed (Figs. F.6 and F.7).

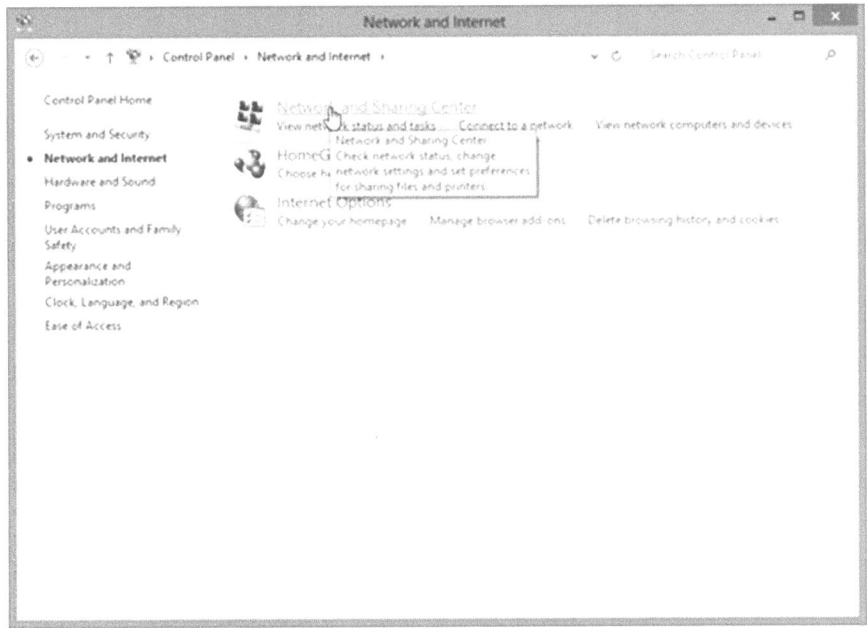

Fig. F.6 Windows 8 Network and Sharing Center screen (Vixen)

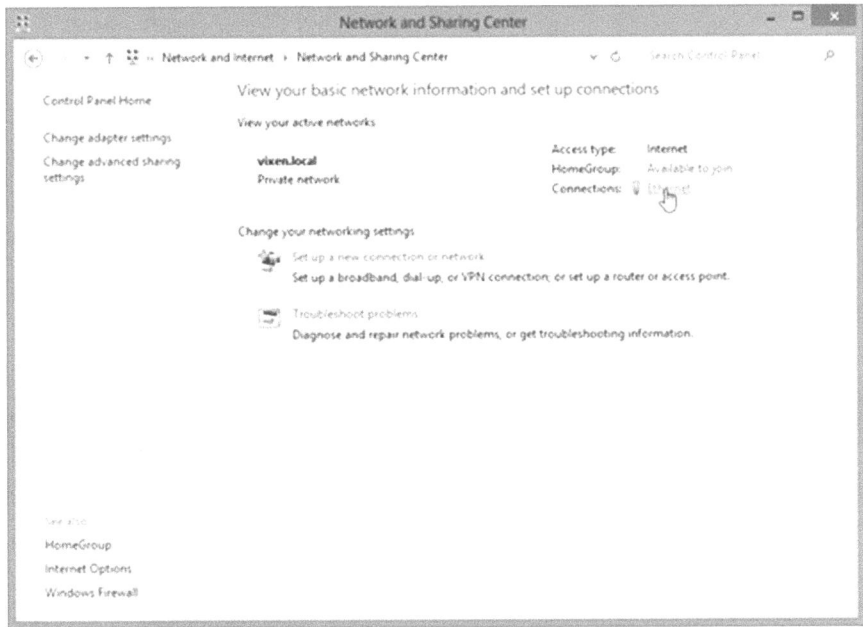

Fig. F.7 Windows 8 Network and Sharing screen 2 (Vixen)

Click on "Ethernet" to go to "Ethernet Status". Click the Properties button to display the "Ethernet Properties" dialog box (Fig. F.8).

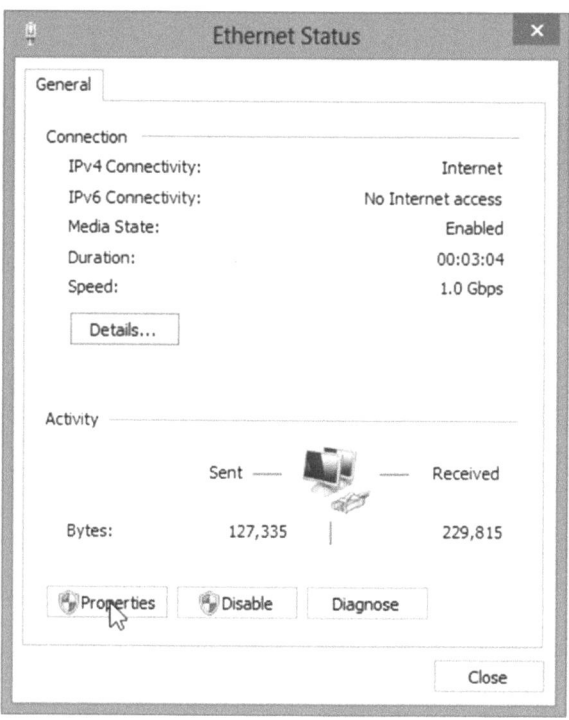

Fig. F.8 Windows 8 Ethernet Status screen (Vixen)

In the window of "This connection uses the following items", scroll down the cursor to "Internet Protocol version 4 (TCP/IPv4)" to select. Click the Properties button in the lower right of the dialog box (Figs. F.9 and F.10).

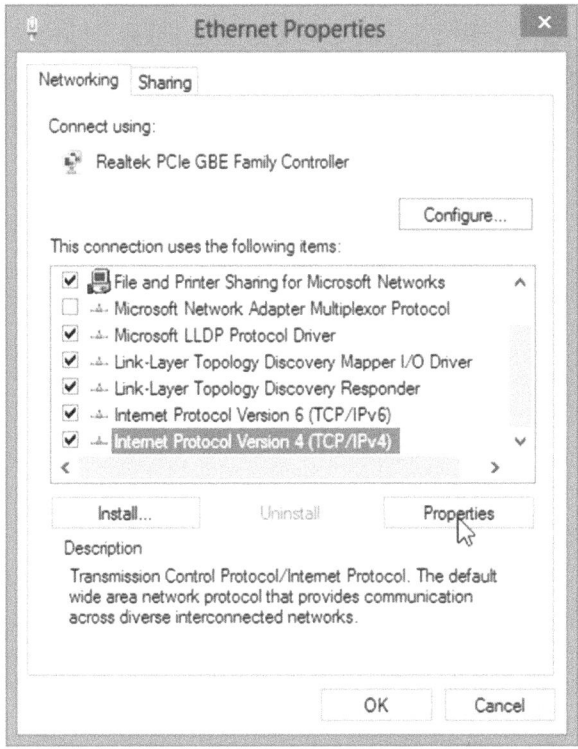

Fig. F.9 Windows 8 Ethernet properties screen (Vixen)

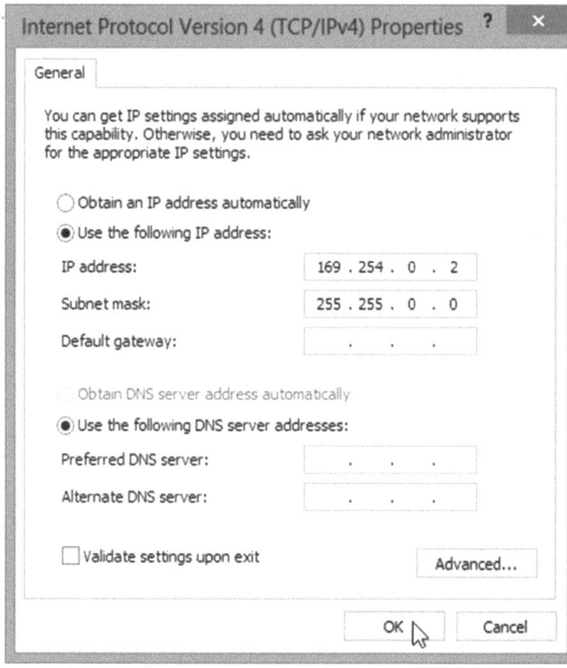

Fig. F.10 Windows 8 Ethernet Properties screen 2 (Vixen)

"Internet Protocol Version 4 (TCP/IPv4) Properties" is displayed. Mark check boxes for "Use the following IP Address" and enter the IP address and Subnet Mask numbers as follows:

IP Address: 169.254.0.2
Subnet Mask: 255.255.0.0

Click the OK button and the settings are completed. Go forward to 6 in the procedure on How to Install the Update Program.

Procedure in Windows 7

Click the Start button (or Windows logo in the bottom left of the screen) to display the Start menu. Select "Control Panel" and click on it to display (Figs. F.11, F.12, and F.13).

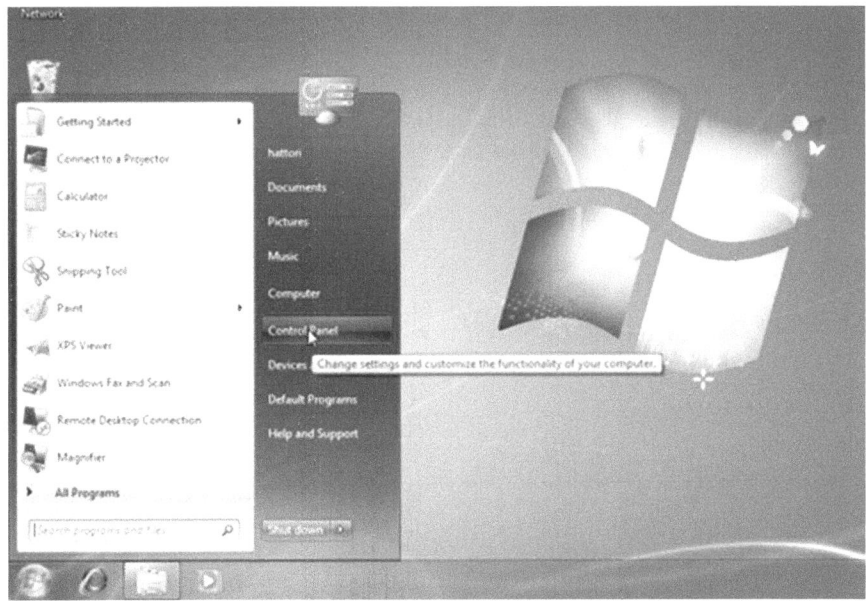

Fig. F.11 Windows 7 Start screen (Vixen)

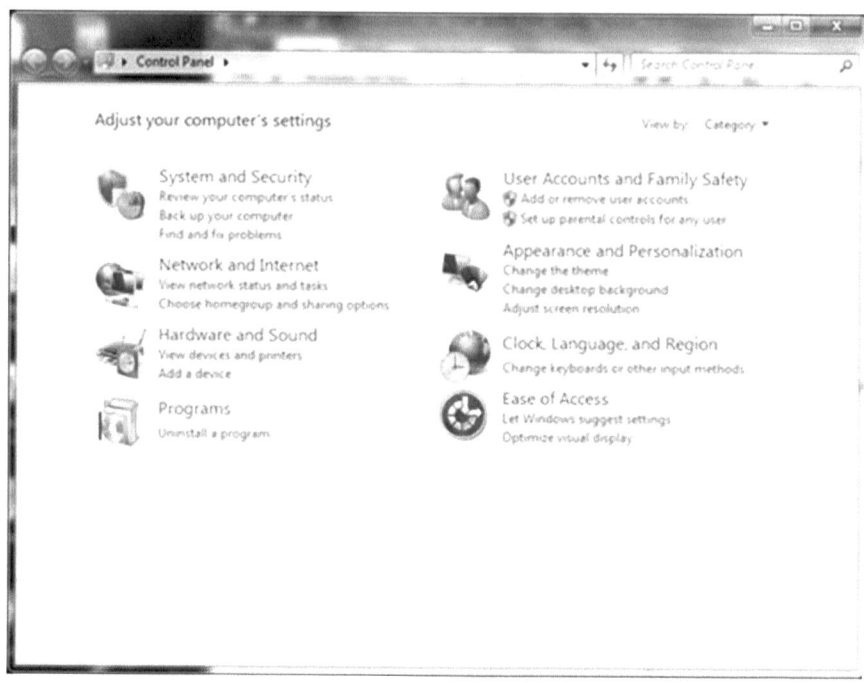

Fig. F.12 Windows 7 Control Panel (Vixen)

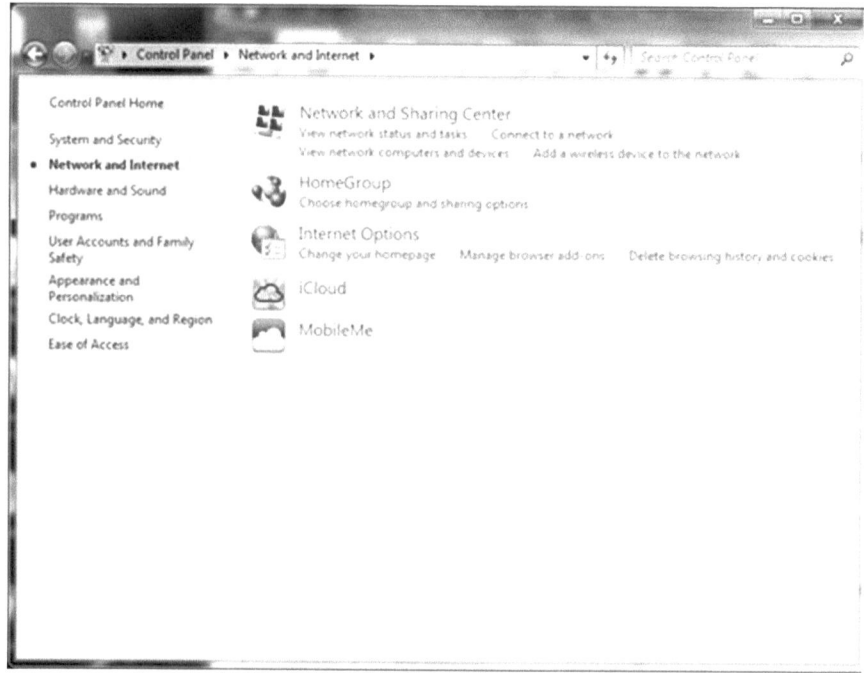

Fig. F.13 Windows 7 Network and Internet screen (Vixen)

Click on "Network and Internet" to display and click on "Network and Sharing Center" in it (Fig. F.14).

Fig. F.14 Windows 7 Local Area Connection Status screen (Vixen)

"Network and Sharing Center" is displayed and Click on "Local Area Connection" in it (Fig. F.15).

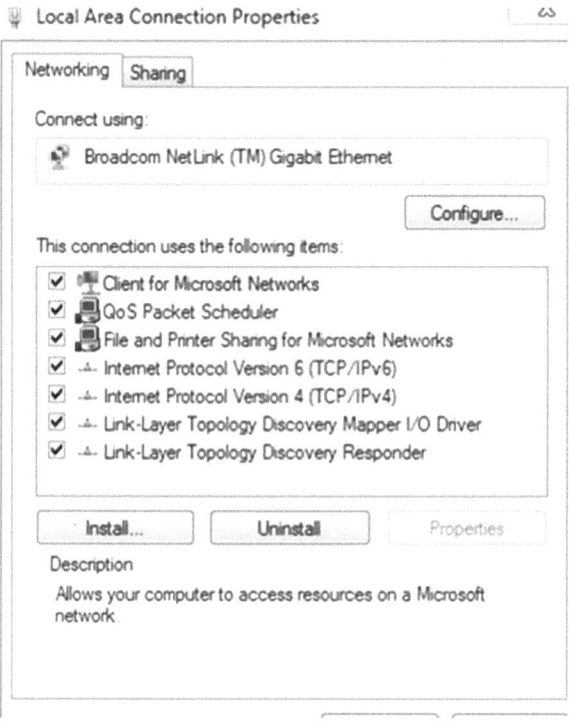

Fig. F.15 Windows 7 Local Area Connection Properties screen (Vixen)

Click the Properties button in the "Local Area Connection" dialog box
(Fig. F.16).

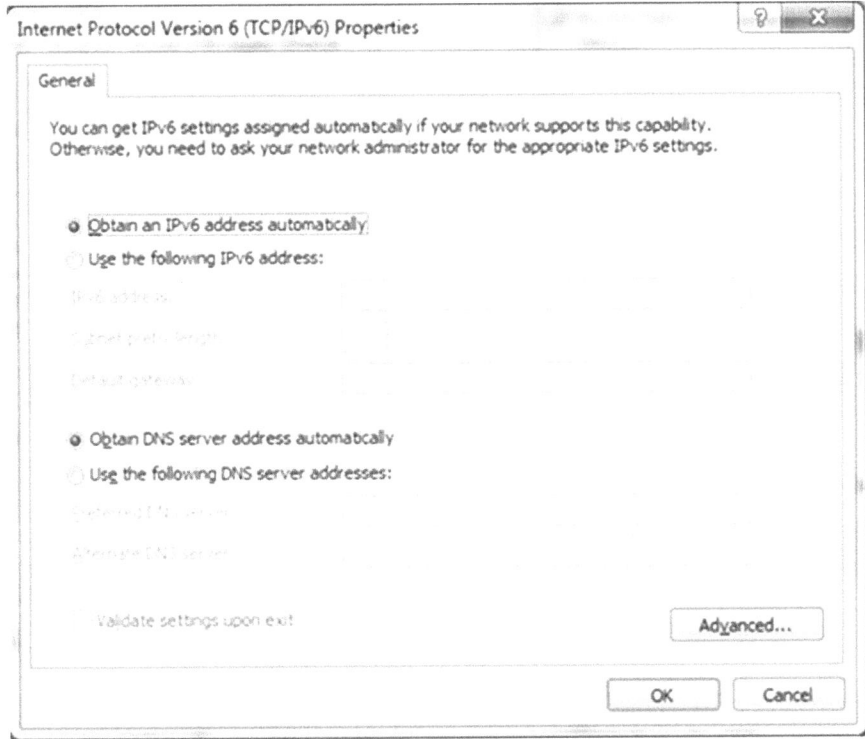

Fig. F.16 Windows 7 Internet Protocol screen (Vixen)

In the window of "This connection uses the following items", scroll down the cursor to "Internet Protocol version 4 (TCP/IPv4)" to select. Click the Properties button in the lower right of the dialog box (Fig. F.17).

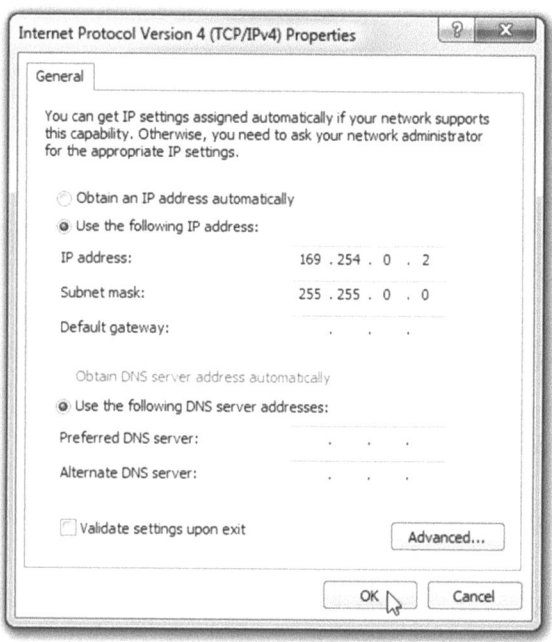

Fig. F.17 Windows 7 Internet Protocol screen cont'd (Vixen)

"Internet Protocol Version 4 (TCP/IPv4) Properties" is displayed. Mark check boxes for "Use the following IP Address" and enter the IP address and Subnet Mask numbers as follows:

IP Address: 169.254.0.2
Subnet Mask: 255.255.0.0

Click the OK button and the settings are completed. Go forward to 6 in the procedure on How to Install the Update Program.

Procedure in Windows Vista

Click the Start button (or Windows logo in the bottom left of the screen) to display
the Start menu (Fig. F.18).

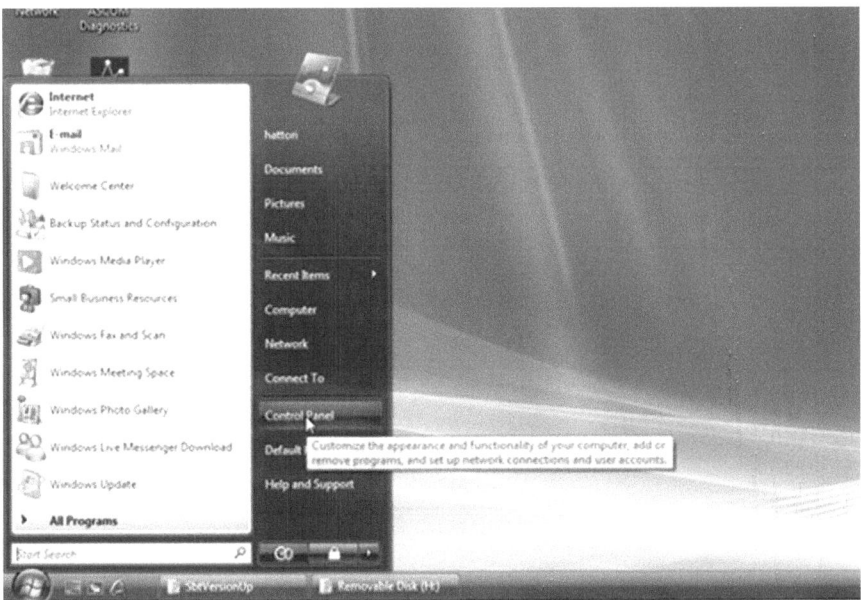

Fig. F.18 Windows Vista Start screen (Vixen)

Click on "Network" to display and click the Properties button in it (Fig. F.19).

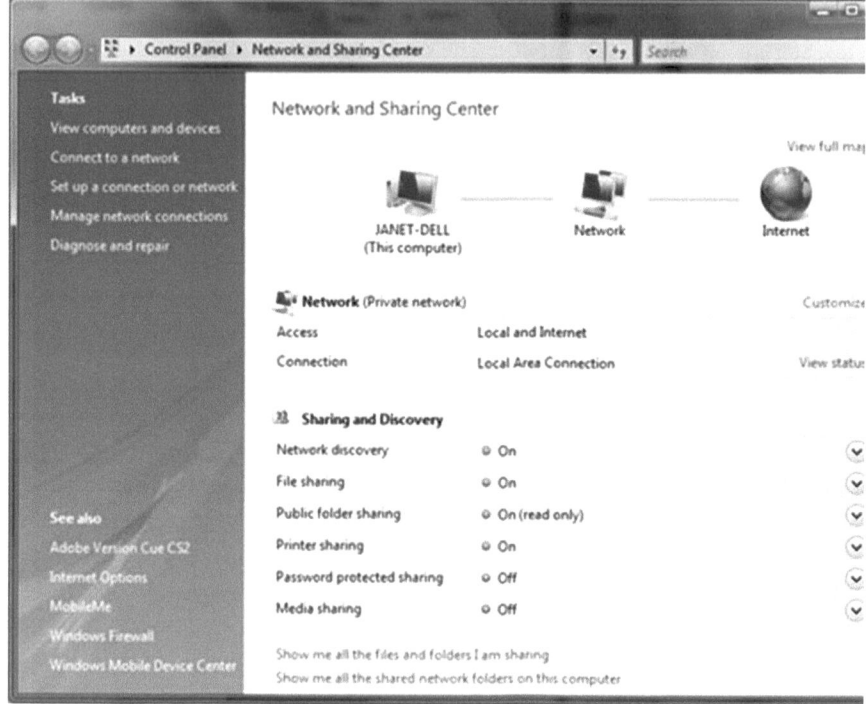

Fig. F.19 Windows Vista Network and Sharing Center screen (Vixen)

"Network and Sharing Center" is displayed and Click on "Manage Network Connections" in the right column (Fig. F.20).

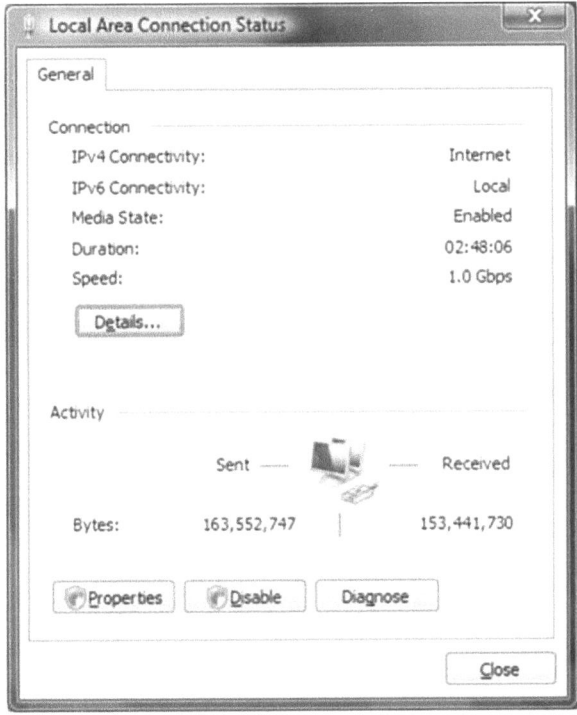

Fig. F.20 Windows Vista Local Area Connection Status screen (Vixen)

Right-click on "Local Area Connection" to display and click the Properties button in the displayed menu (Fig. F.21).

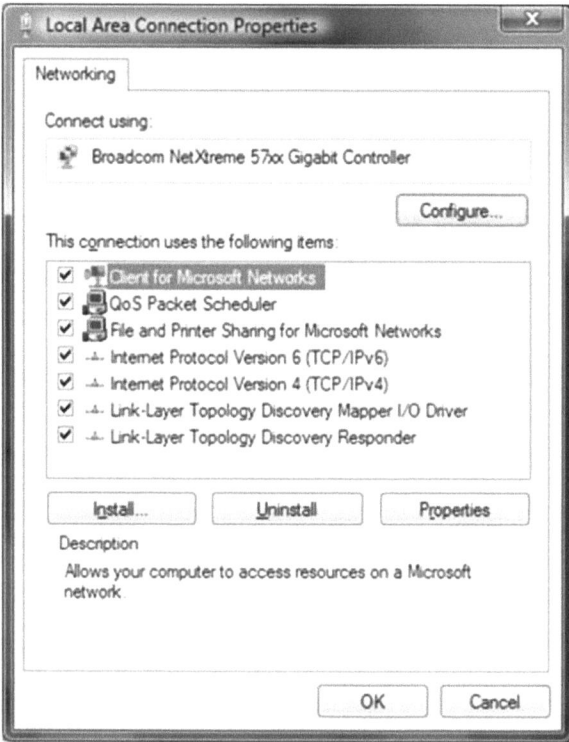

Fig. F.21 Windows Vista Local Area Connection Properties screen (Vixen)

※Click the Continue button if the dialog box below appears on the screen
(Fig. F.22).

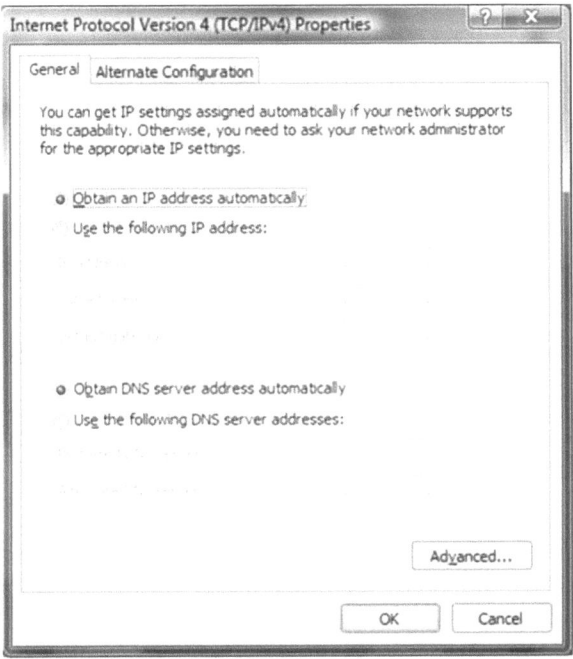

Fig. F.22 Windows Vista Internet Protocol properties screen (Vixen)

In the windows of "This connection uses the following items", scroll down the cursor to "Internet Protocol version 4 (TCP/IPv4)" to select. Click the properties button in the lower right of the dialog box (Fig. F.23).

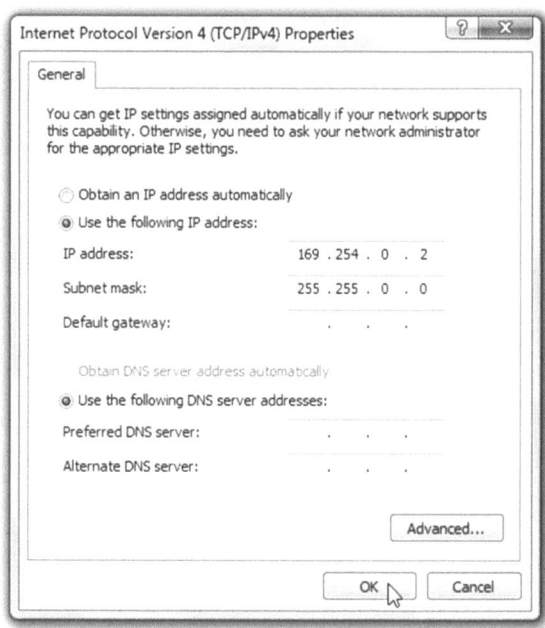

Fig. F.23 Windows Vista Internet Protocol properties screen cont'd (Vixen)

"Internet Protocol Version 4 (TCP/IPv4) Properties" is displayed. Check the checkbox at "Use the following IP Address" and enter the IP address and Subnet Mask numbers as follows:

IP Address: 169.254.0.2
Subnet Mask: 255.255.0.0

Click the OK button and the settings are completed. Go forward to 6 in the procedure on How to Install the Update Program.

Procedure in Windows XP

Click the Start button (or Windows logo in the bottom left of the screen) to display the Start menu. Select "Control Panel" and click on it to display (Figs. F.24 and F.25).

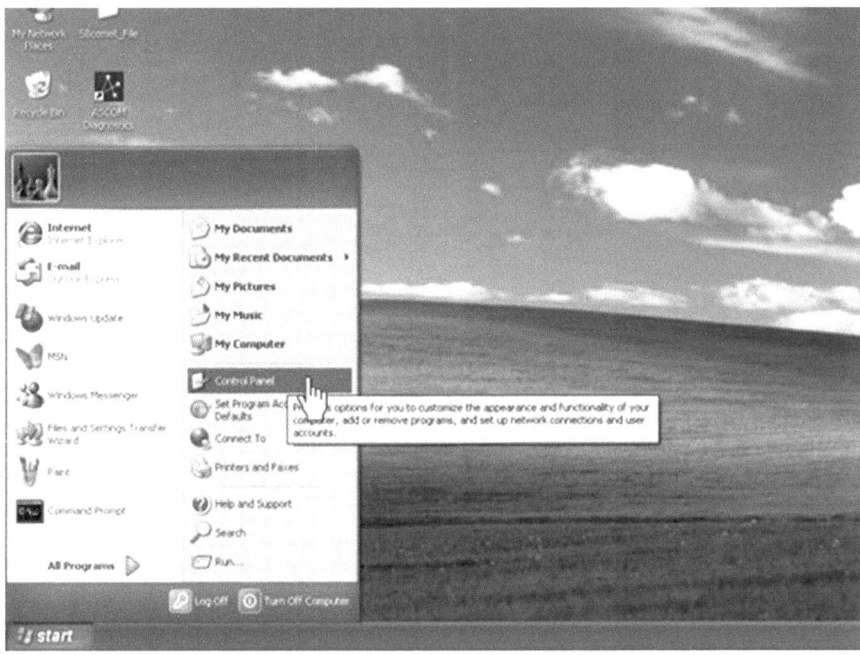

Fig. F.24 Windows XP Start screen (Vixen)

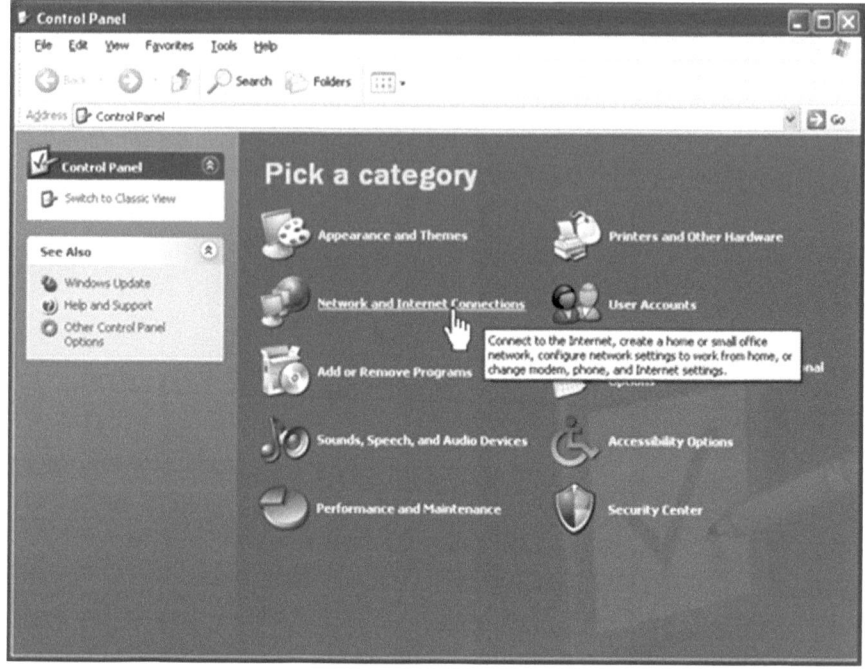

Fig. F.25 Windows XP Control Panel (Vixen)

In the "Control Panel" menu, click on "Network and Internet" to display (Fig. F.26).

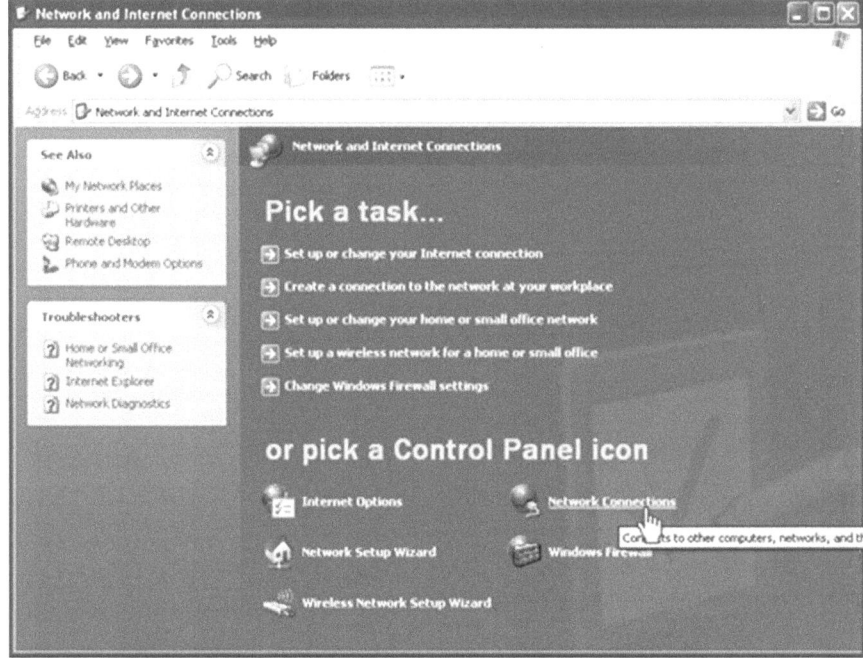

Fig. F.26 Windows XP Network and Internet Connections screen (Vixen)

Click on "Network Connections" (Fig. F.27).

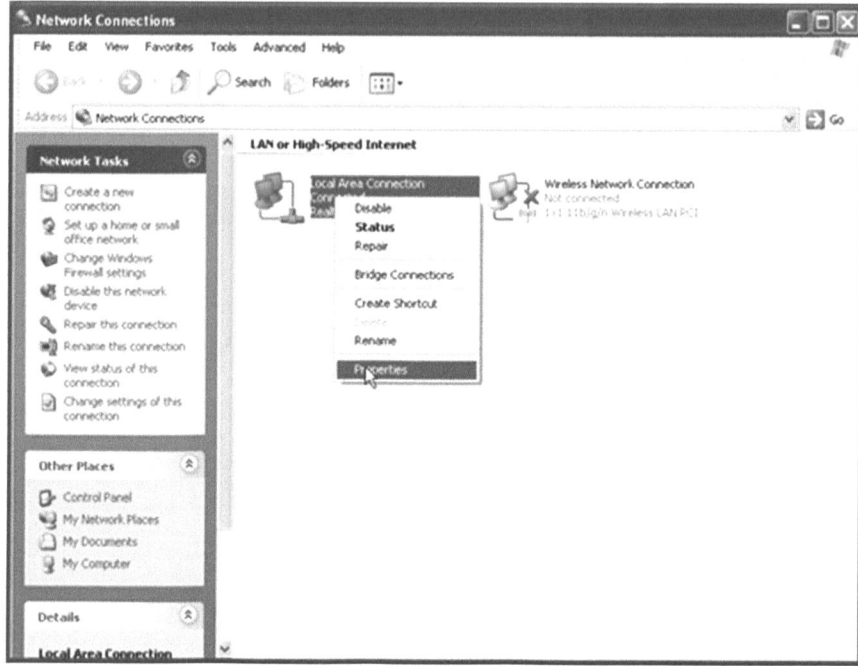

Fig. F.27 Windows XP Network Connections screen (Vixen)

Right-click on "Local Area Connection" to display and click the Properties button in the displayed menu (Fig. F.28).

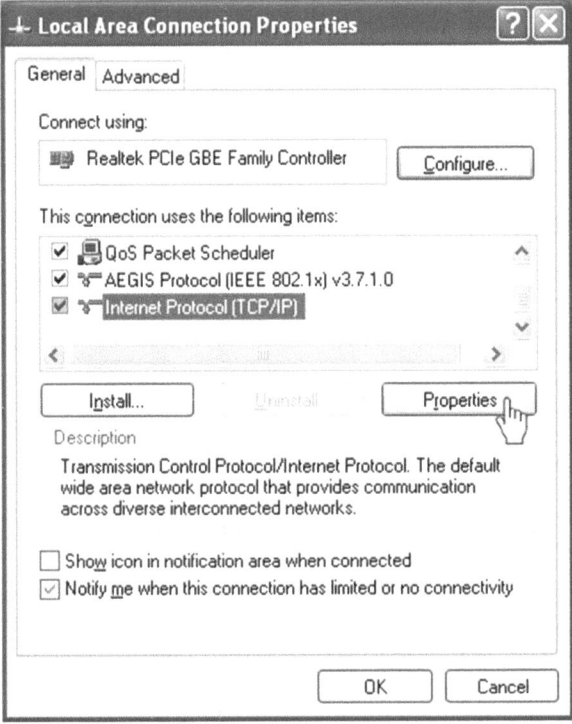

Fig. F.28 Windows XP Local Area Connection Properties screen (Vixen)

In the window of "This connection uses the following items", scroll down the cursor to "Internet Protocol version 4 (TCP/IPv4)" to select. Click the Properties button in the lower right of the dialog box (Fig. F.29).

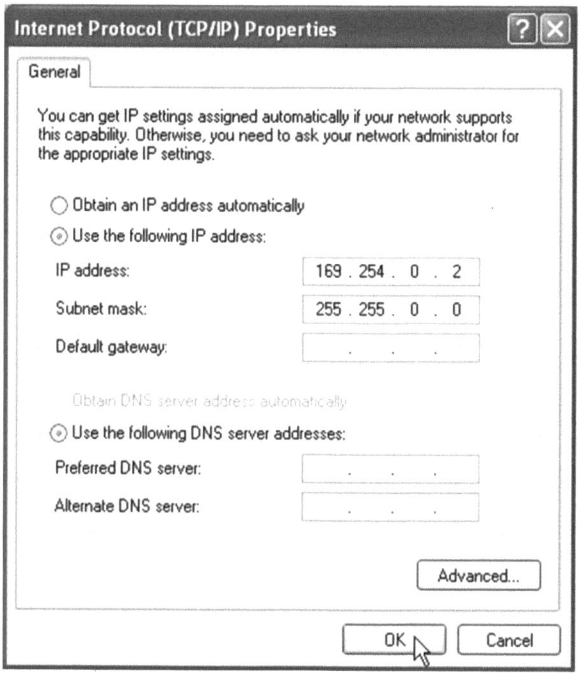

Fig. F.29 Windows XP Internet Protocol Properties screen (Vixen)

"Internet Protocol Version 4 (TCP/IPv4) Properties" is displayed. Mask check boxes for "Use the following IP Address" and enter the IP address and Subnet Mask numbers as follows:

IP Address: 169.254.0.2
Subnet Mask: 255.255.0.0

Click the OK button and the settings are completed. Go forward to 6 in the procedure on How to Install the Update Program (Figs. F.30, F.31, and F.32).

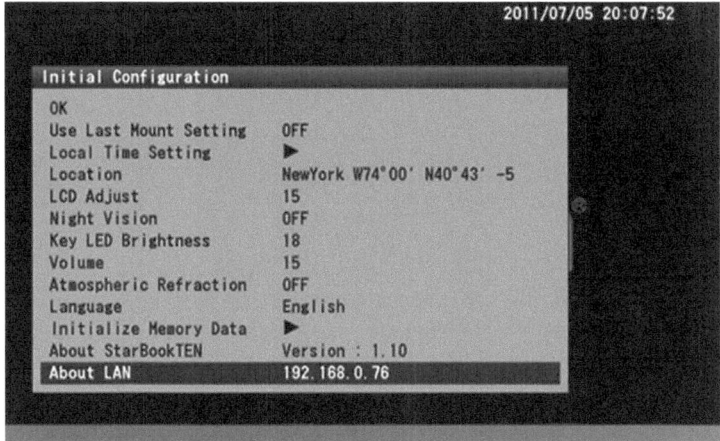

Fig. F.30 Vixen Star Book TEN Intial Configuration screen (Vixen)

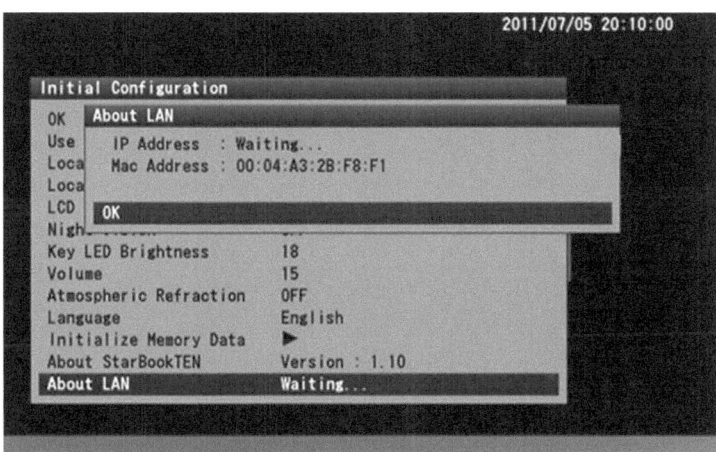

Fig. F.31 Vixen Star Book TEN About LAN screen (Vixen)

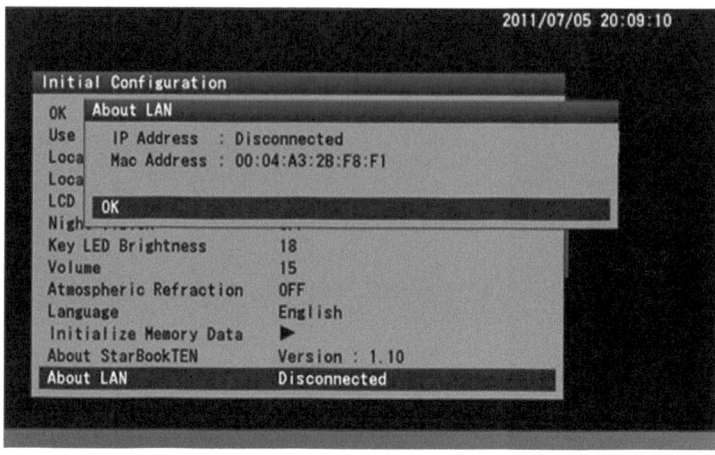

Fig. F.32 Vixen Star Book TEN About LAN cont'd (Vixen)

Wait a period of one minute. The STAR BOOK TEN and PC automatically acquire information on networking.

Note:

Never unplug the STAR BOOK TEN during the communication with the PC, and do not disconnect the LAN crossover cable. The update program may be destroyed and the STAR BOOK TEN may fail to reboot if the power supply is turned off during this process.

※The IP address may differ if the STAR BOOK TEN is not connected to the PC by means of one to one connection, or depending on settings on the PC.

If the connection to the PC does not finish, the message "Warning..." will be displayed. Or, if the connection ends in failure, the message "Disconnected" will be displayed. If this is the case, wait a short time and make sure that the LAN crossover cable is connected securely (Figs. F.33 and F.34).

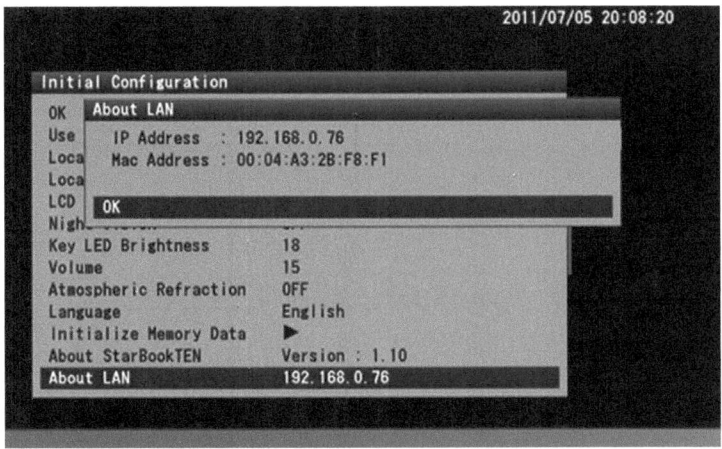

Fig. F.33 Vixen Star Book TEN About LAN cont'd (Vixen)

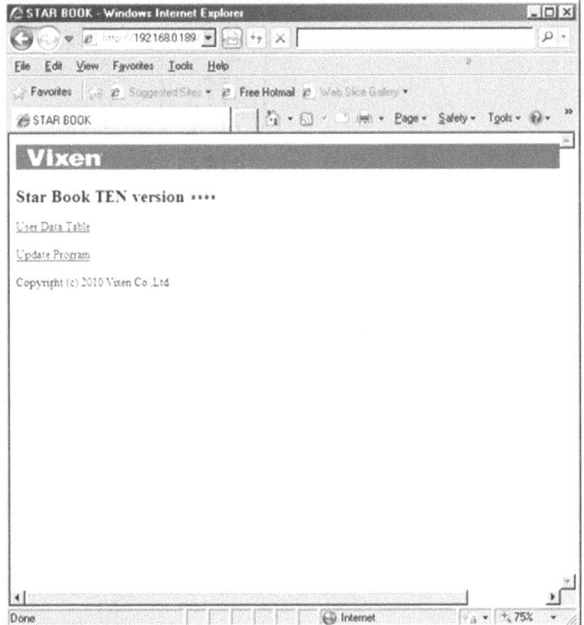

Fig. F.34 Internet Explorer IP Address entry (Vixen)

Open Internet Explorer and enter the IP address you obtained in step 6 into the address bar to display the entry page.

Example: If the IP address is 192.168.0.189, put http;//192.168.0.189 on the address bar.

※Change the setting to have the address bar appear if it is hidden on your PC.

The entry page shown on the left appears. Click on "Update Program" in the entry page (Fig. F.35).

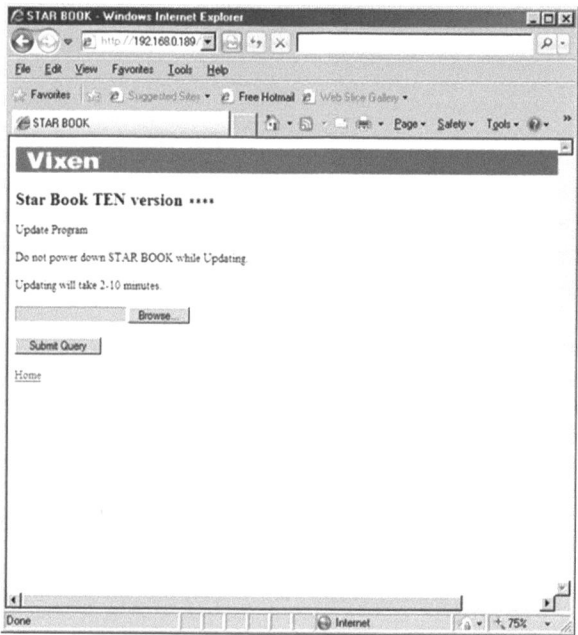

Fig. F.35 Internet Explorer Star Book TEN version screen (Vixen)

The "Program Update" page appears. Click the Browse button (Fig. F.36).

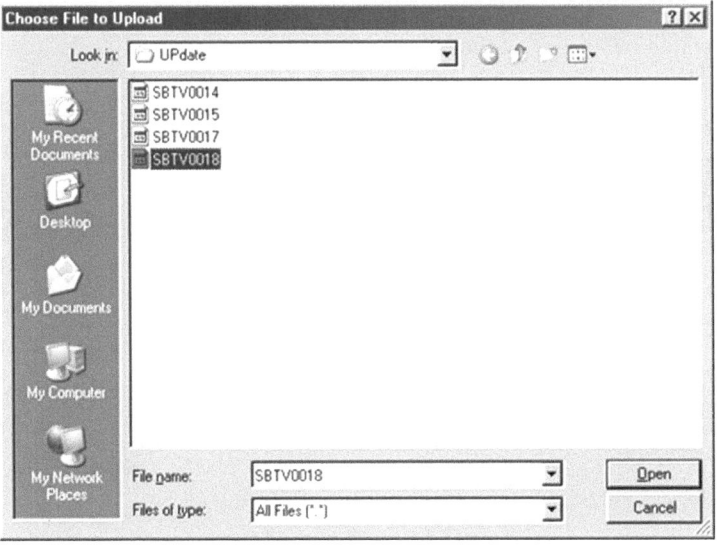

Fig. F.36 Program Update screen (Vixen)

Designate the file folder that contains the update program file you downloaded.

Select the update program file and click the Open button.

Example of File Name: SBTV0110.bin (Update Program Ver.1.10) (Fig. F.37)

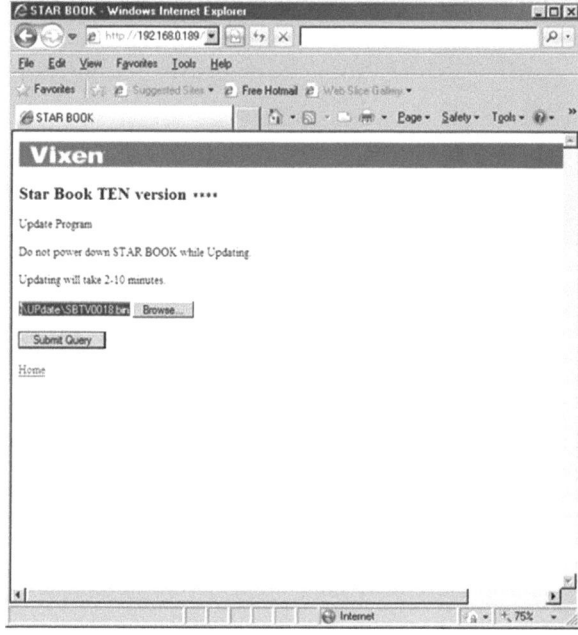

Fig. F.37 Update Program Selection example (Vixen)

Click the Submit Query button to send (Figs. F.38, F.39, and F.40).

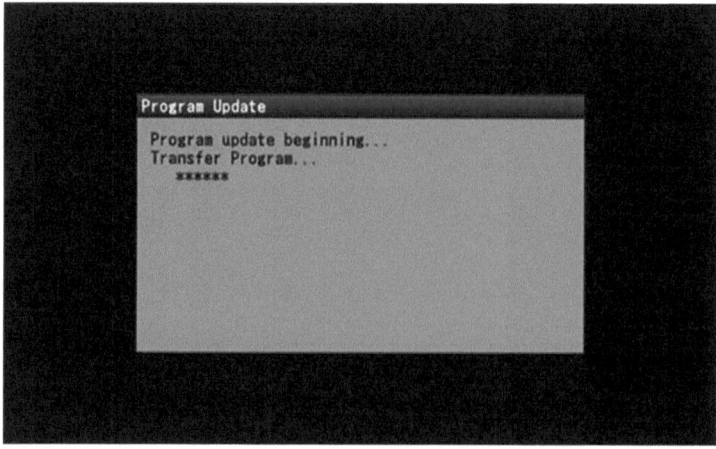

Fig. F.38 Program Update screen 1 (Vixen)

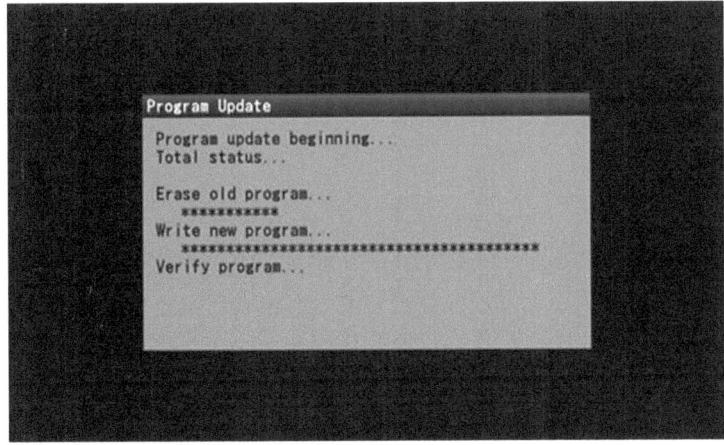

Fig. F.39 Program Update screen 2 (Vixen)

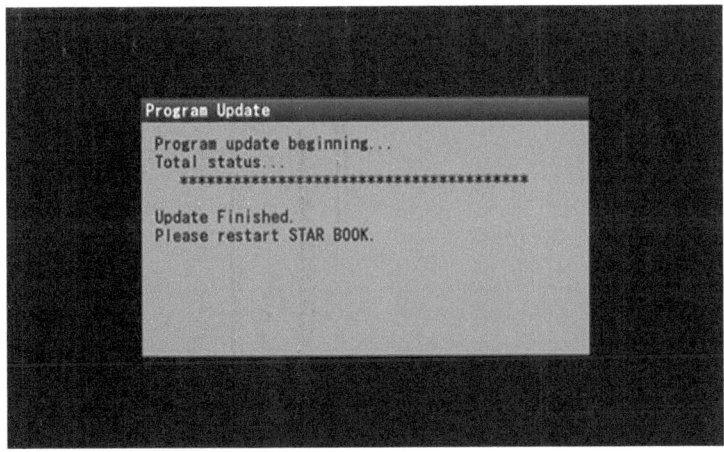

Fig. F.40 Program Update screen 3 (Vixen)

The update of the STAR BOOK TEN starts. The message on the left is displayed on the screen and the progress of the update is indicated with a graphic bar (Fig. F.41).

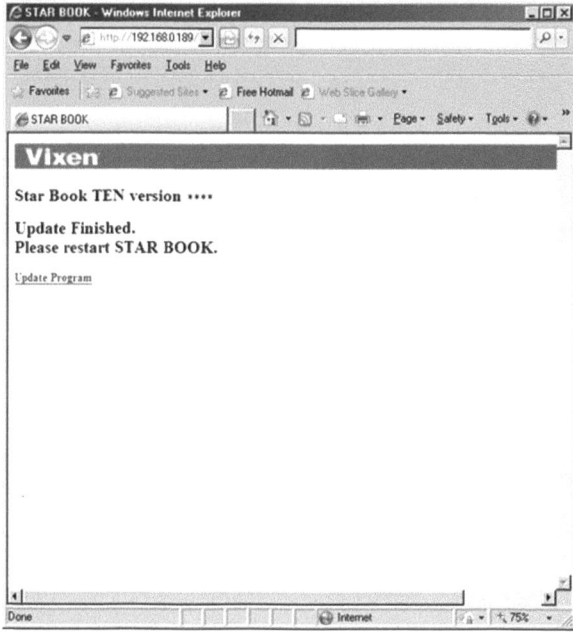

Fig. F.41 Vixen Star Book TEN Update Program screen (Vixen)

At the end of the update, the message indicating a successful update appears on the screens of both the PC and STAR BOOK TEN. Turn off and reboot the STAR BOOK TEN (Fig. F.42).

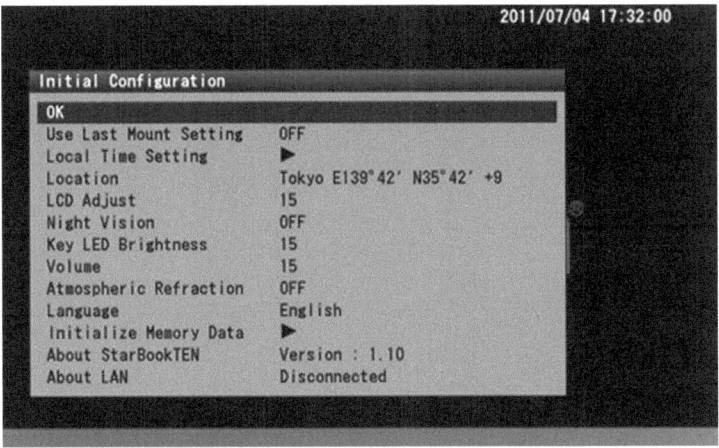

Fig. F.42 Vixen Star Book TEN reboot (Vixen)

Now the update has been installed successfully. In the "Initial Configuration" menu, go to "About STAR BOOK TEN" and confirm the version updated. Restore the settings on the PC if changed.

Appendix G

Star Book Updates

As of this writing, the most recent firmware build for the Star Book is Version 2.6 of May 26, 2014. The firmware version of the Star Book controller used for the writing of this book is version 2.2.

In most cases, it is not necessary for the user to continually update the firmware of the Star Book with every iteration. The author's recommendation is if the Star Book is performing to the needs of the user, leave it alone. Performing the firmware update is not for the faint of heart and the computing neophyte. It is highly recommended that the user has the update performed by the Vixen dealer or by Vixen itself.

All updates can only be performed using a PC computer. This author uses an Apple iMac running iOS 10.9.5, and as a result did not and cannot perform any updates to the Star Book or Star Book TEN firmware. The author cannot verify or validate the Vixen procedures as presented.

Firmware Updates History

Improving Slewing Accuracy to the Moon

It improves the pointing accuracy of automatic go-to slewing to the Moon with minute position calculations of the Moon.

Program Correction

When choosing "AltAz" mode in the chart setting, it happened possibly that the accurate pointing of the Moon and planets was disturbed.

When ascending celestial objects below horizon reached less than 1° in altitude, there was an error that the "minus (−)" was not displayed.

© Springer International Publishing Switzerland 2016   205
J.L. Chen, A. Chen, *The Vixen Star Book User Guide*, The Patrick Moore
Practical Astronomy Series, DOI 10.1007/978-3-319-21593-8

There was an error that the stop button did not work while the mount moved to the home position.

With installation of this update program, all the above errors are corrected. We sincerely apologize for any inconvenience this has caused you.

Program Error Correction

An error that the setting is not saved when you choose "Save the Setting" to hold the setting is corrected.

In the setting of the constellation name, an error that the constellation name is turned to OFF regardless of ON setting is corrected.

Addition of target objects to the "Star" database in the Object Menu.

The following fixed stars become available for display from the Object Menu.

Rigel (Orion)

Becrux (Crux)

Hadar (Centaurus)

Program correction

This update is to remedy erratic behaviors which arise at tracking of the SX and SXD mounts in the setting of the southern hemisphere only. There is no difference between the Ver.2.2 and the Ver.2.1 in the tracking performance as long as the mounts are used at the setting of the northern hemisphere.

Program correction

In the previous version of the programs, there was a malfunction that a volume setting of the sound could not be saved. You can retain the current volume setting with installation of the update program. We apologize for causing you trouble in this matter.

Other improvement

A feeling of the operation buttons becomes smoother and the behaviors of the motors in tracking are refined.

Change in the description of Pluto

The description of Pluto has been changed to "Pluto (dwarf planet)" in accordance with the classification change.

Program correction

The troubles that are listed below were found in STAR BOOK.

The malfunctions are corrected by installing the program here, We apologize for the inconvenience.

The automatic go-to system does not finish when using an SXD Equatorial Mount.

The tracking is unstable when using an SXD Equatorial Mount.

PEC Function

An accurate auto-tracking of celestial objects is possible.

Go-To Function—Comets

The go-to system for comets can be implemented by entering the comet's orbital elements available to the general public. The method used in this function is the same as in the simple go-to function for comets released in the past.

*Go-To Speed Control (*1)*

The speed of celestial go-to system can be reduced to about 1/2 of the original speed. It reduces the noise produced when go-to function is performed.

*1...The function is available in SXD mounts and specific SX mounts which have a remodeled motor control board. The function is not available in SXW/SXC mounts which have the original motor control board. (Not displayed in the menu.)

The names of prominent bright fixed stars and Messier objects can be displayed on the screen. The manner of displaying the names can be selected from "Always ON", "Always OFF" and steps at zooming.

Change the motor speed in the initial setting screen

The motor speed can be changed in the initial setting screen. It enables you to set the suitable motor speed separately for when adjusting the telescope direction by slow motion, when adjusting the finderscope, or when observing the landscape.

The Display Function for Confirmation Dialogs

Now a dialog that asks whether automatic go-to system can be started is displayed after you select the go-to system. If you click No, the chart mode can be switched to the scope mode without executing the automatic go-to process.

Home Position Function

When observation is finished, the direction of optical tube can be changed back to the initial position by the simple procedure. For the models without Decl. clamps, this function is helpful in resetting the telescope position for the next observation.

Stars with magnitude up to 7.0 are displayed.

The data of the stars with magnitude up to 7.0 are stored in STAR BOOK. It had only displayed the stars up to 6.0 in the past, but the stars with 7.0 magnitude can now be displayed by updating to this version.

Spanish language supported

The display in Spanish is available.

Turning OFF the backlight

Can be turned off during observation.

Auto-Guider Function

Can be used for auto-guider adapters such as AGA-1, SBIG, and SBIG ST series.

Backlash Correction Function

It is the function that corrects the backlash in SX equatorial mounts. Manual operation becomes easy by installing and running this function.

Increasing the motion speed on screen

The speed of the motion of screen like scrolling speed became much faster the past versions (Build 1.1 and before). Operating from the screen is now easier than before.

"Stellar Navigator Ver.8" and "Stellar Gear" Supported

SX equatorial mounts will be supported by a control software of Astro Arts, "Stellar Navigator Ver. 8" and "Stellar Gear."

※Contact Astro Arts for the details of the software "Stellar Navigator Ver. 8" and "Stellar Gear."

Adjustment of the time/date in a leap year

The program that displays time and date in a leap year can be adjusted to display correctly.

LCD adjustment setting is available

The settings for "contrast" and "brightness" can be saved. After adjusting the parts on the LCD screen, execute "save the settings" command to save each of the settings. The changes will be in effect when the power is turned off and re-started.

Update Procedure

The following is taken from the Vixen Co. Japan website (www.vixen.co.jp). The author is not responsible for its contents. The photos of the screen were also taken from the website, and may not be readable in this book. Again, the author stresses before attempting any update, please refer to the Vixen website for the most up-to-date procedures and the most clear pictures.

The following exemption is posted on the Vixen website. This author does not take any responsibility for any damage experienced by a user of the Star Book TEN following these instructions:

In case of any damage experienced by a customer when installing or using this software or any charge sent by a third party to a customer due to the damage, Vixen or their dealer do not take responsibility unless the customer is not responsible and Vixen or their dealer was negligent.

Warning for Update

The Sphinx mounts with original motor control board

The go-to speed varying function can not be used in the SX equatorial mounts (SXW and SXC mounts) equipped with the original motor control board.

※It's possible to install the program, but the functions that are unavailable will not be displayed in the menu.

How to check the motor control board on your Sphinx mount

It can be checked in the initial setup menu right after the power switch is turned to ON. Select "About STAR BOOK" with the arrow keys (↑/↓) and press "select." Either of the followings is displayed on the upper right of the screen,

T...(T1a3, etc.) It means that the mount has the original motor control board.
S...(S2a7, etc.) It is the mount with the remodeled motor control board (new board).

Other Warning

The version upgrades are provided as part of the package. A function cannot be installed separately.

For the users of SX equatorial mounts: If you are using STAR BOOK Ver 1.2 and earlier versions, this update would reset the current settings of observational location and LCD intensity. Please re-enter the current settings.

eCos is used in part of this software

Based on the license issued by users can obtain the source code used in the eCos part Please contact Vixen for the detail.

Exemption

In case of any damage experienced by a customer when installing or using this software or any charge sent by a third party to a customer due to the damage, the company or dealers do not take responsibility unless the customer is not responsible or the company was intentional or negligent.

Update Procedure

Before You Begin

Please use new batteries or AC adapter or stabilized power supply when you upgrade the current version of your STAR BOOK. Also, use a stable power outlet for the PC to which data is transferred.

There may be a problem with re-starting STAR BOOK if the power is turned off during an upgrading process.

In such cases, contact a dealer to request a repair.(additional fee)

(Please note that the setting may be initialized during the repair.)

The network setting of the PC may have to be changed to update the program. Please save the settings so that they can be retrieved after the upgrade.

The current settings in STAR BOOK (such as Latitude/Longitude of observing location) may be initialized during a version update. Please take a note or save the settings before you start the update process.

Requirements

STAR BOOK

Power outlet for STAR BOOK (the outlet on which SX/SXD equatorial mount can be operated)

PC to which the data is forwarded (LAN card for a model without a LAN terminal)

※ Please see your PC (or LAN card) manual about the detail of LAN cards.

Crossover LAN cable

Applicable OS Environment of the PC

Applicable OS environment:	Microsoft Windows®XP Home Edition, XP Professional, Vista, 7 Home Edition, 7 Professional, 7 Ultimate, 8, 8Pro, 8.1, 8.1Pro
Web Browser:	Internet Explorer version 5.0 or higher
CPU:	Pentium®II 400 MHz or better
RAM:	256 MB or more
LAN:	10BASE-T/100BASE-T

※Depending on the OS used, necessary information may not be displayed properly. If this is the case, update your OS to the latest version using Windows Update.

※Microsoft and Windows are registered trademarks of Microsoft Corporation in the US and some Countries.

All product and company names mentioned in the instructions are trademarks or registered trademarks of the respective companies.

Download Procedure

When connecting STAR BOOK to your PC

※Click below if you are connecting STAR BOOK directly to LAN environment or a router.

IMPORTANT NOTE—For the customers who use a STAR BOOK in German language:

Be sure to choose English language in the initial setting menu before you start updating. Choosing German language in your update will meet with failure.

Download one of the STAR BOOK update programs before connecting to a PC.

Connect the STAR BOOK and the PC with the LAN crossover cable.

※As for a connection of the LAN cable to the PC, please refer to the instructions for the PC.

Turn on the STAR BOOK.

Turn on the PC.

Wait for a period of one minute. The STAR BOOK and PC acquire information on networking automatically (Figs. G.1, G.2, and G.3).

In the System Menu of the STAR BOOK, scroll down the cursor to select "About STAR BOOK" and press Select key to enter. Confirm that the following IP Address and Subnet Mask numbers are displayed.

- IP Address: 169.254.a.b (a, b are arbitrary numbers)
- SubNet Mask: 255.255.0.0

Double click on the downloaded program file (Such as update_v21.exe) to run the program.

※ In Windows Vista and Windows 7 a dialog box shown on the left (User Account Control) comes out for warning. Select "Allow" (A) I trust this program to proceed.

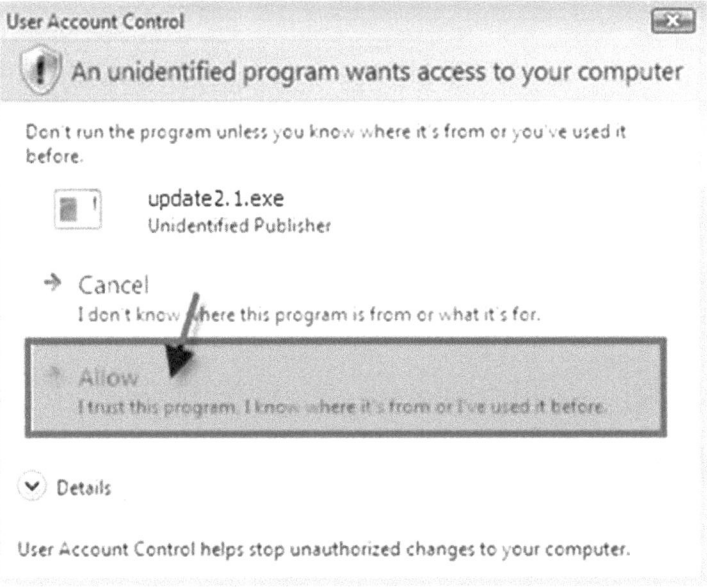

Fig. G.1 Star Book and PC networking screen 1 (Vixen)

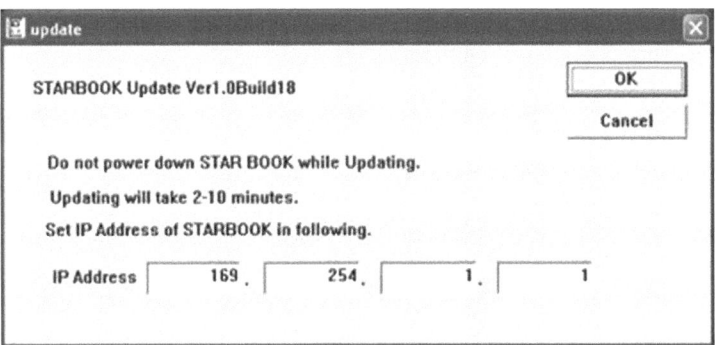

Fig. G.2 Star Book and PC Networking screen 2 (Vixen)

Fig. G.3 Star Book and PC Networking screen 3 (Vixen)

Enter the IP Address as you obtained at 6. Click on the OK button.

<Ex.> For 169.254.1.1 → 169.254.1.1

Wait for a few minutes until the message on the left appears on the screen.

Turn off the STAR BOOK to reboot. Turn on the STAR BOOK again. Scroll down the cursor to go into "About STAR BOOK". Confirm that the version is updated.

Your update is finished successfully. (Go back to the previous settings of the PC if you changed for the update.)

If the procedure above does not work

※If your update is unsuccessful, follow the directions below to change settings of the PC. (For Windows XP, Windows Vista and Windows 7, administrator authorization is required.)

Procedure in Windows 8

Right-click on the background in the start up screen. Click "All apps" in the bottom right of the screen and display every application.

※ For Windows 8.1

Click the down arrow mark in the bottom left of the screen to open "apps" (Figs. G.4, G.5, and G.6).

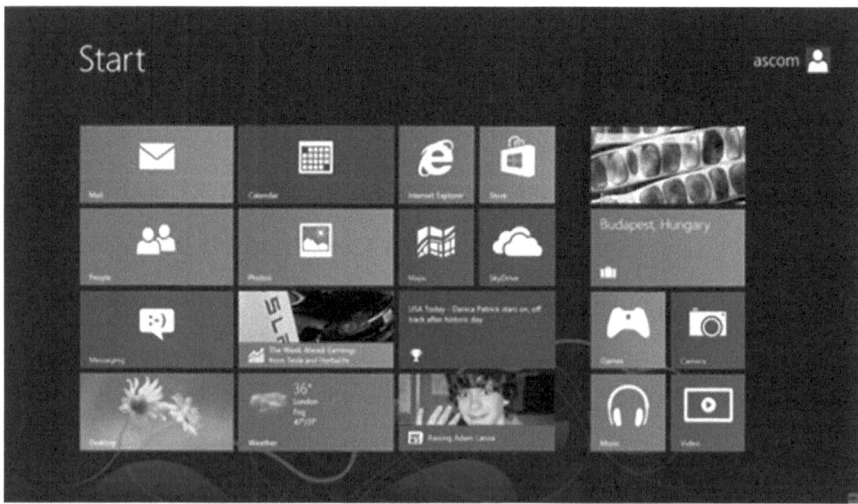

Fig. G.4 Windows 8 Start screen (Vixen)

Fig. G.5 Windows 8 Start screen cont'd (Vixen)

Fig. G.6 Windows 8 Apps screen (Vixen)

Select "Control Panel" and go to "Network and Internet" (Figs. G.7 and G.8).

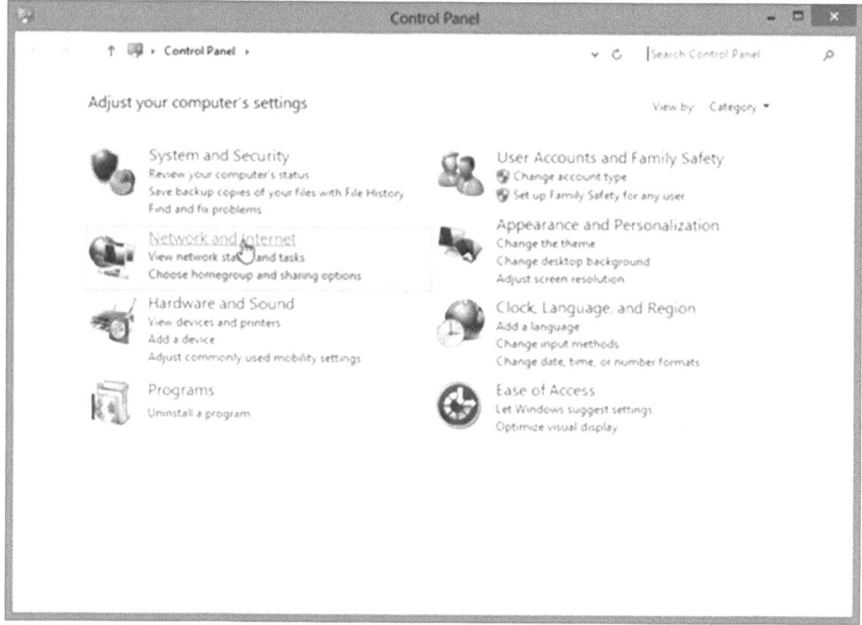

Fig. G.7 Windows 8 Control Panel screen (Vixen)

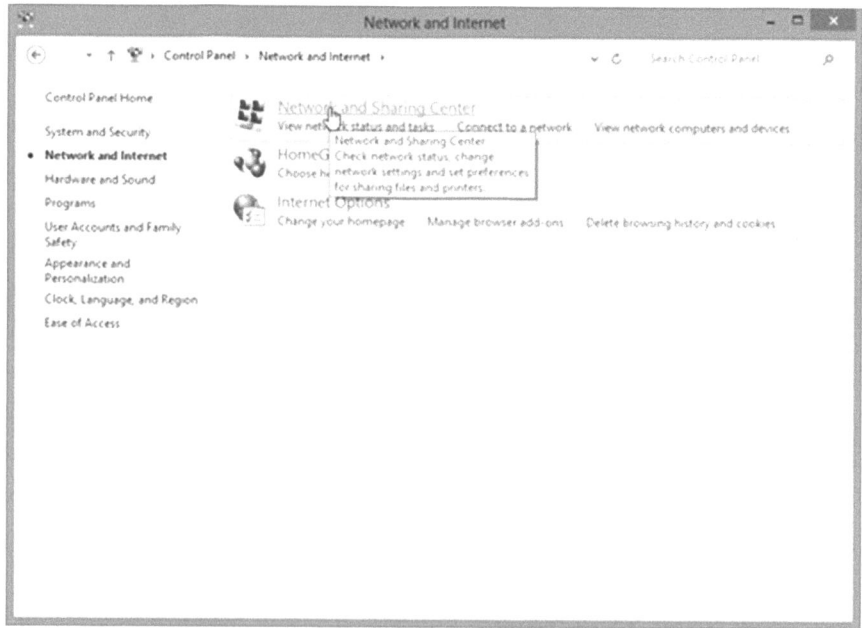

Fig. G.8 Windows 8 Network and Internet screen (Vixen)

Click on "Network and Sharing Center" The "View your basic network information and set up connections" screen is displayed (Fig. G.9).

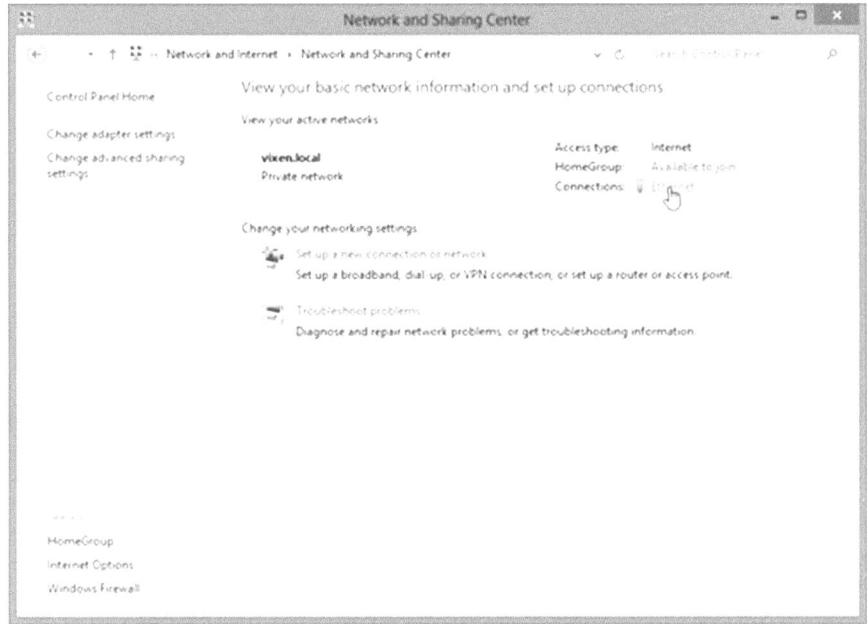

Fig. G.9 Windows 8 Network and Sharing Center screen (Vixen)

Click on "Ethernet" to go to "Ethernet Status". Click the Properties button to display the "Ethernet Properties" dialog box (Fig. G.10).

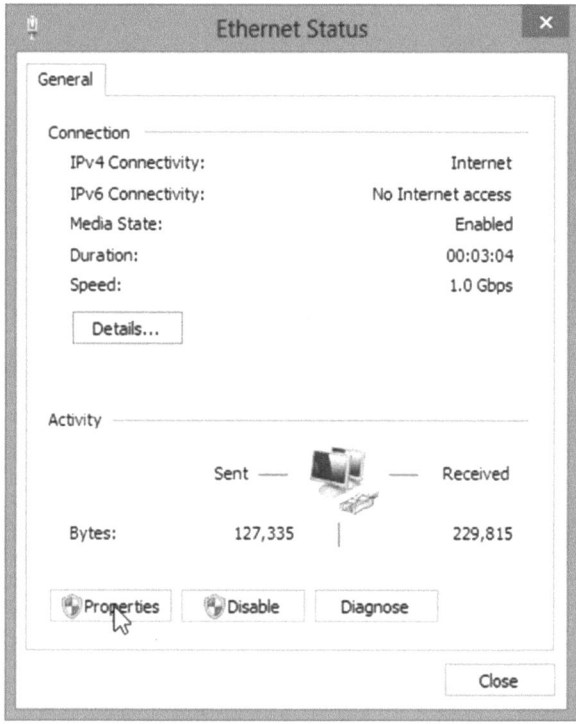

Fig. G.10 Windows 8 Ethernet Status screen (Vixen)

In the window of "This connection uses the following items", scroll down the cursor to "Internet Protocol version 4 (TCP/IPv4)" to select. Click the Properties button in the lower right of the dialog box (Figs. G.10 and G.11).

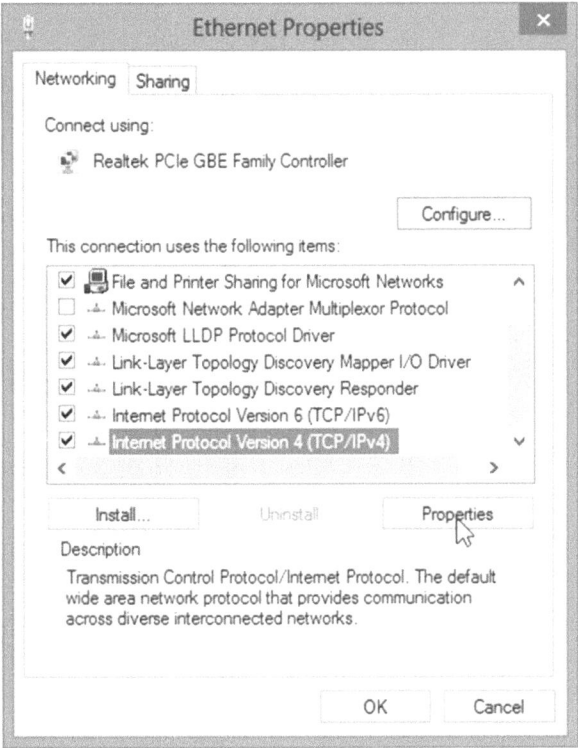

Fig. G.11 Windows 8 Ethernet Properties screen (Vixen)

"Internet Protocol Version 4 (TCP/IPv4) Properties" is displayed. Mark check boxes for "Use the following IP Address" and enter the IP address and Subnet Mask numbers as follows:

IP Address: 169.254.1.2
Subnet Mask: 255.255.0.0

Click the OK button and the settings are completed. Go forward to 6 in the procedure on How to Install the Update Program.

Procedure in Windows 7

Click the Start button (or Windows logo in the bottom left of the screen) to display the Start menu. Select "Control Panel" and click on it to display (Figs. G.12 and G.13).

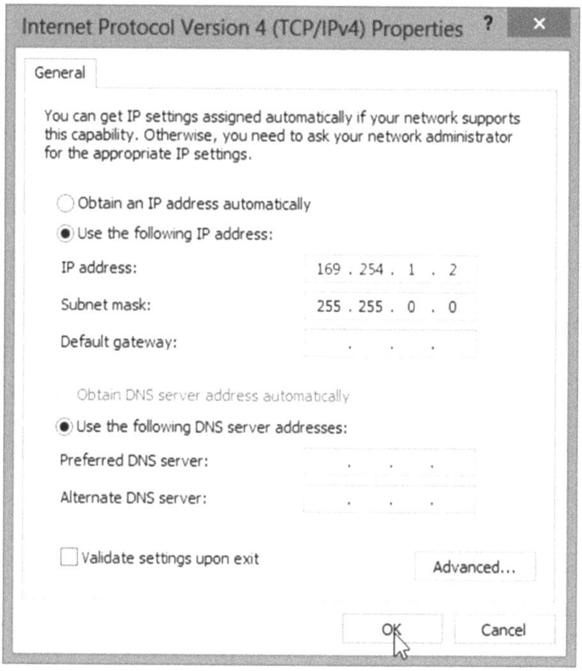

Fig. G.12 Windows 7 Internet Protocol screen (Vixen)

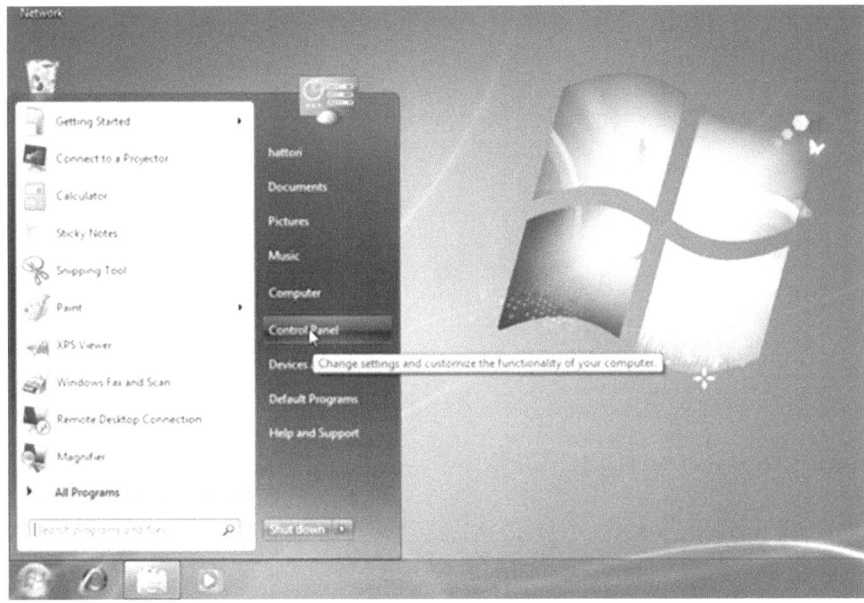

Fig. G.13 Windows 7 Start screen (Vixen)

Click on "Network and Internet" to display and click on "Network and Sharing Center" in it (Fig. G.14 and G.15).

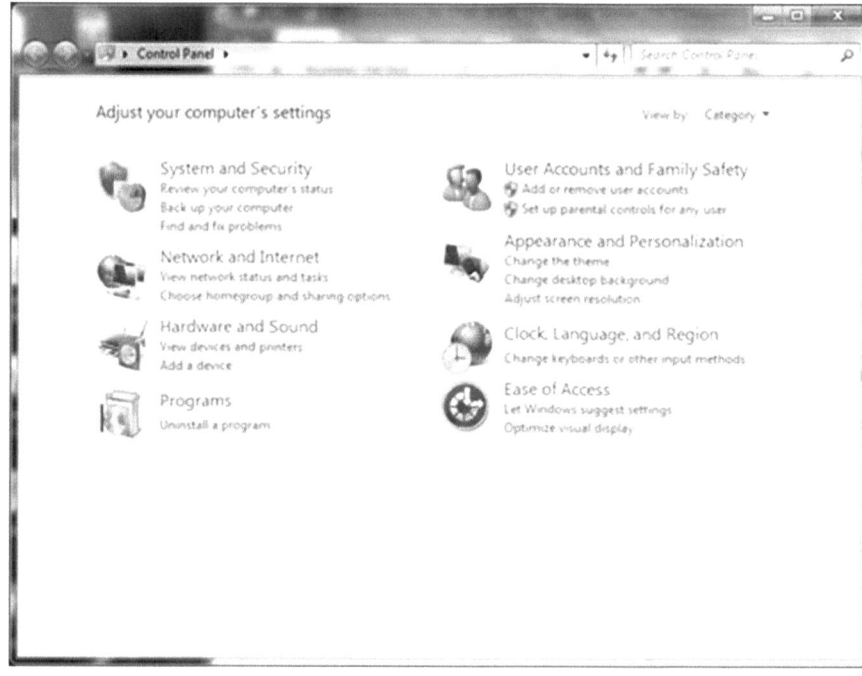

Fig. G.14 Windows 7 Control Panel screen (Vixen)

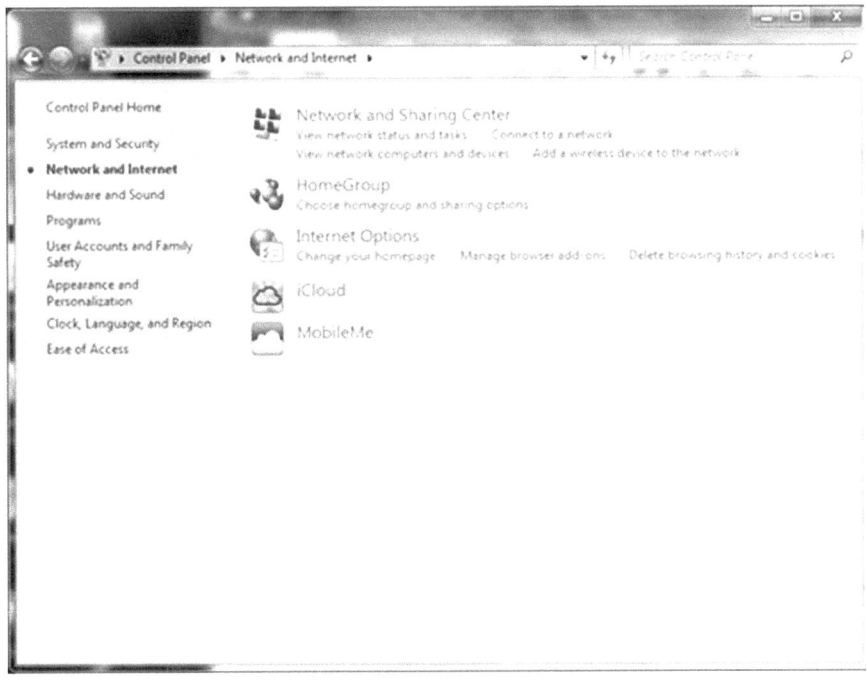

Fig. G.15 Windows 7 Network and Internet screen (Vixen)

"Network and Sharing Center" is displayed and Click on "Local Area Connection" in it (Fig. G.16).

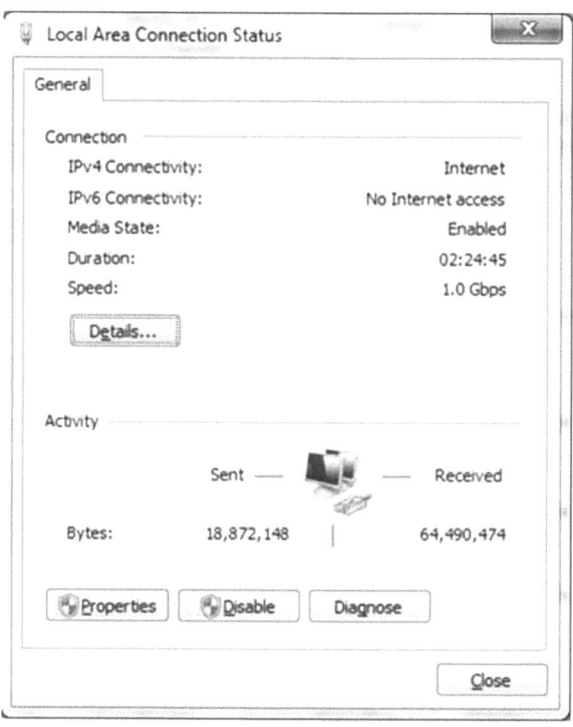

Fig. G.16 Windows 7 Local Area Connection Status screen (Vixen)

Click the Properties button in the "Local Area Connection" dialog box (Fig. G.17).

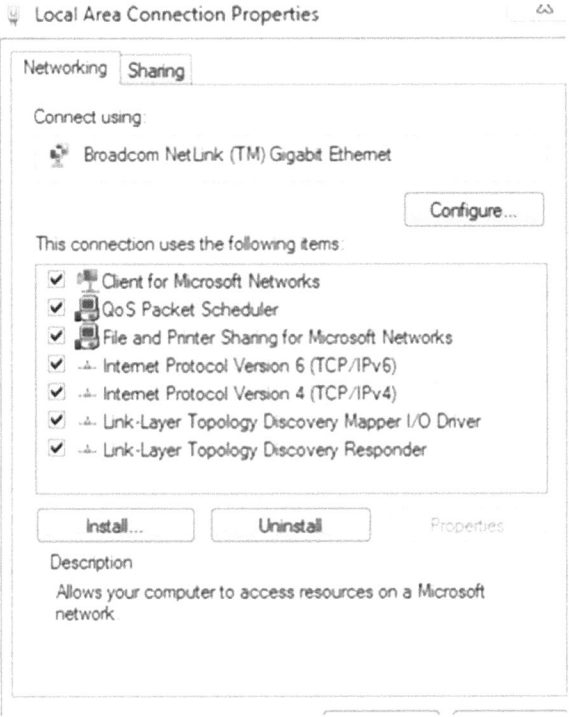

Fig. G.17 Windows 7 Local Area Connection Properties screen (Vixen)

In the window of "This connection uses the following items", scroll down the cursor to "Internet Protocol version 4 (TCP/IPv4)" to select. Click the Properties button in the lower right of the dialog box (Figs. G.18 and G.19).

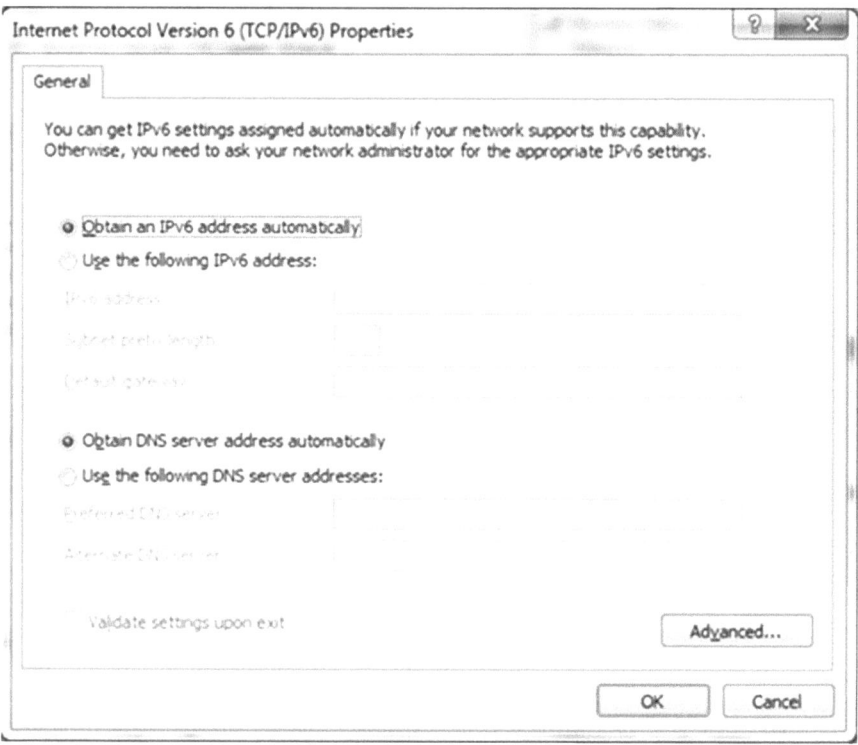

Fig. G.18 Windows 7 Internet Protocol Properties screen (Vixen)

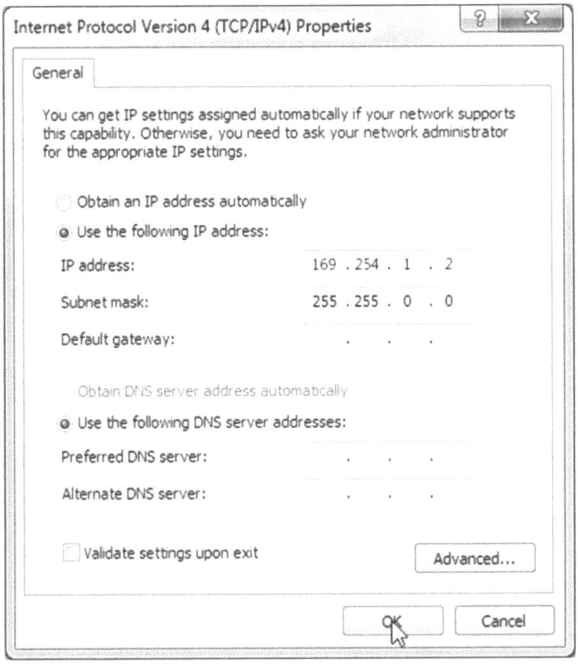

Fig. G.19 Windows 7 Internet Protocol Properties cont'd (Vixen)

"Internet Protocol Version 4 (TCP/IPv4) Properties" is displayed. Mark check boxes for "Use the following IP Address" and enter the IP address and Subnet Mask numbers as follows:

IP Address: 169.254.1.2
Subnet Mask: 255.255.0.0

Click the OK button and the settings are completed. Go forward to 6 in the procedure on How to Install the Update Program.

Click the Start button (or Windows logo in the bottom left of the screen) to display the Start menu (Fig. G.20).

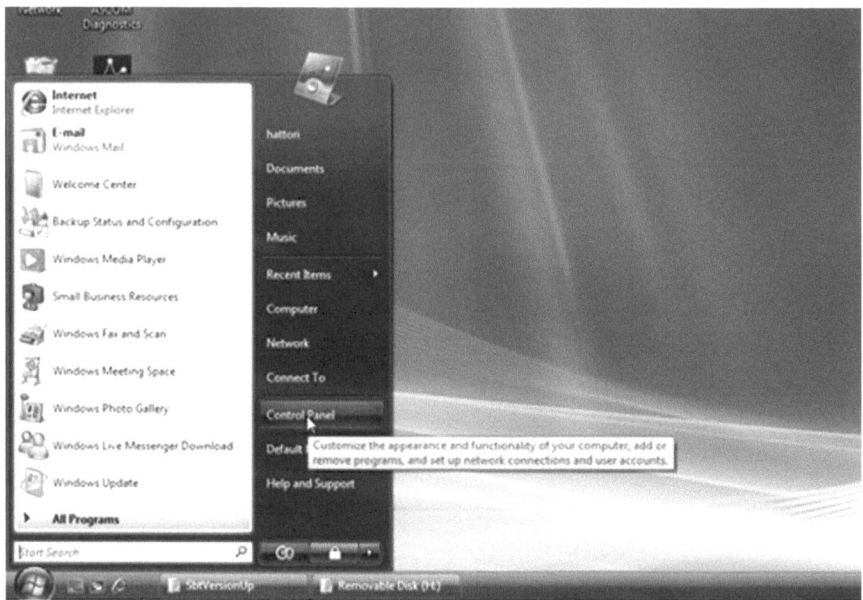

Fig. G.20 Windows Vista Start screen (Vixen)

Click on "Network" to display and click the Properties button in it (Fig. G.21).

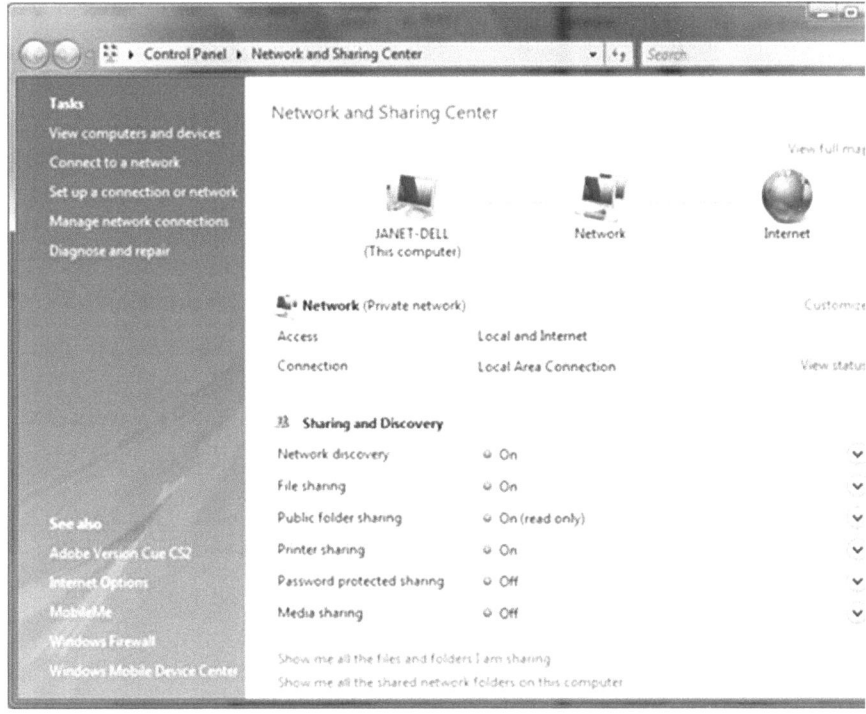

Fig. G.21 Windows Vista Netwrok and Sharing Center screen (Vixen)

"Network and Sharing Center" is displayed and Click on "Manage Network Connections" in the right column (Fig. G.22).

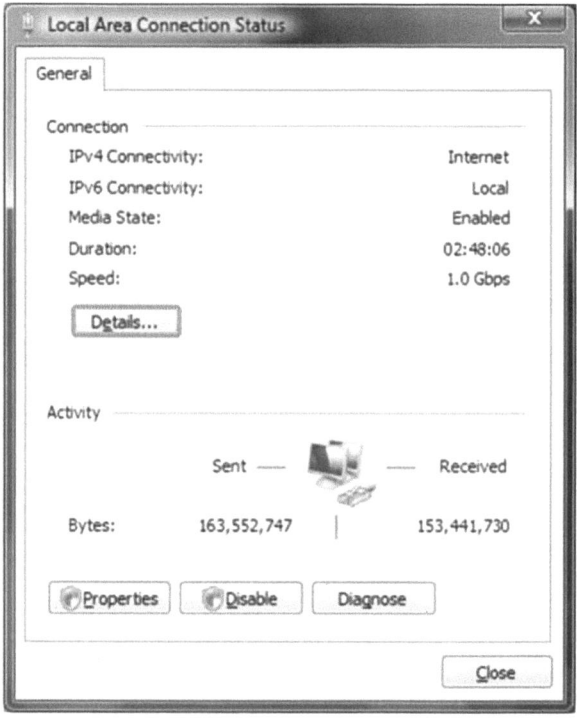

Fig. G.22 Windows Vista Local Area Connection Status screen (Vixen)

Right-click on "Local Area Connection" to display and click the Properties button in the displayed menu (Fig. G.23).

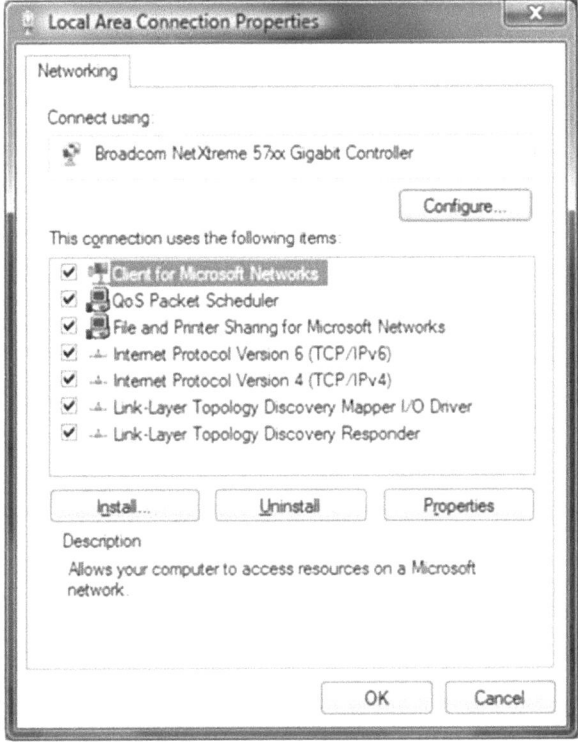

Fig. G.23 Windows Vista Local Area Connection Properties screen (Vixen)

※Click the Continue button if the dialog box below appears on the screen (Fig. G.24).

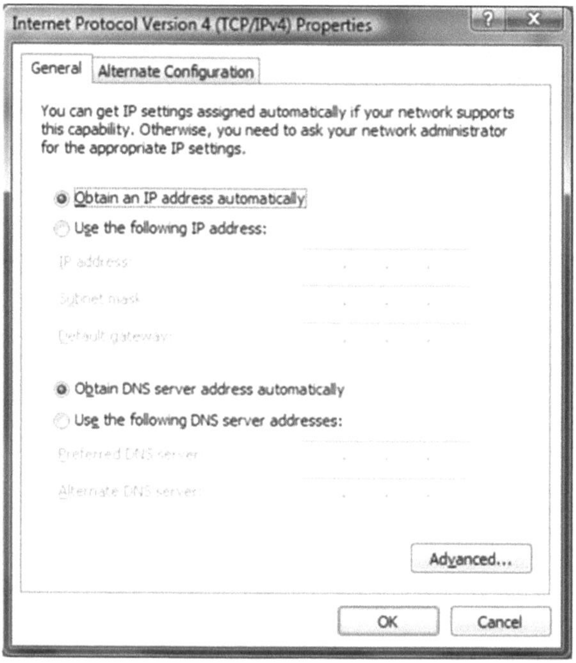

Fig. G.24 Windows Vista Internet Protocol Properties screen (Vixen)

In the windows of "This connection uses the following items", scroll down the cursor to "Internet Protocol version 4 (TCP/IPv4)" to select. Click the properties button in the lower right of the dialog box (Fig. G.25).

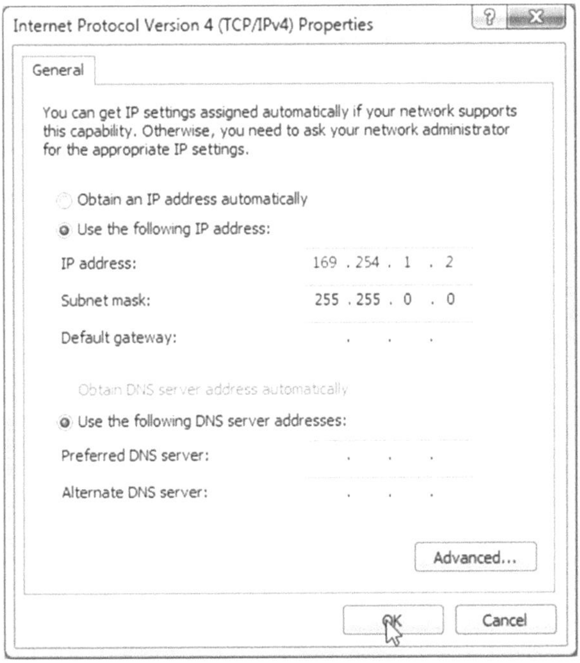

Fig. G.25 Windows Vista Internet Protocol Properties cont'd (Vixen)

"Internet Protocol Version 4 (TCP/IPv4) Properties" is displayed. Check the checkbox at "Use the following IP Address" and enter the IP address and Subnet Mask numbers as follows:

IP Address: 169.254.1.2
Subnet Mask: 255.255.0.0

Click the OK button and the settings are completed. Go forward to 6 in the procedure on How to Install the Update Program.

Click the Start button (or Windows logo in the bottom left of the screen) to display the Start menu. Select "Control Panel" and click on it to display (Fig. G.26).

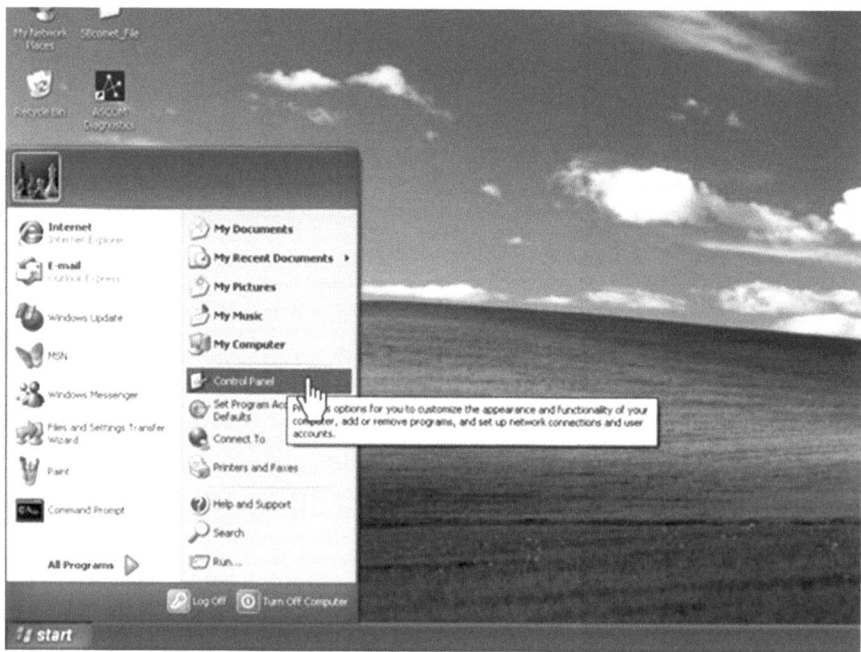

Fig. G.26 Windows XP Start screen (Vixen)

In the "Control Panel" menu, click on "Network and Internet" to display (Fig. G.27).

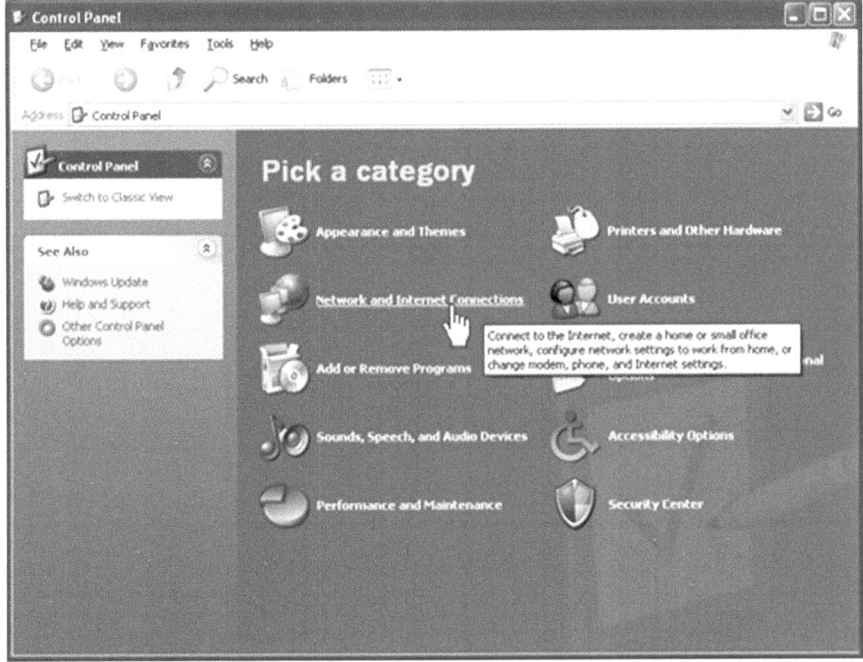

Fig. G.27 Windows XP Control Panel screen (Vixen)

Click on "Network Connections" (Fig. G.28).

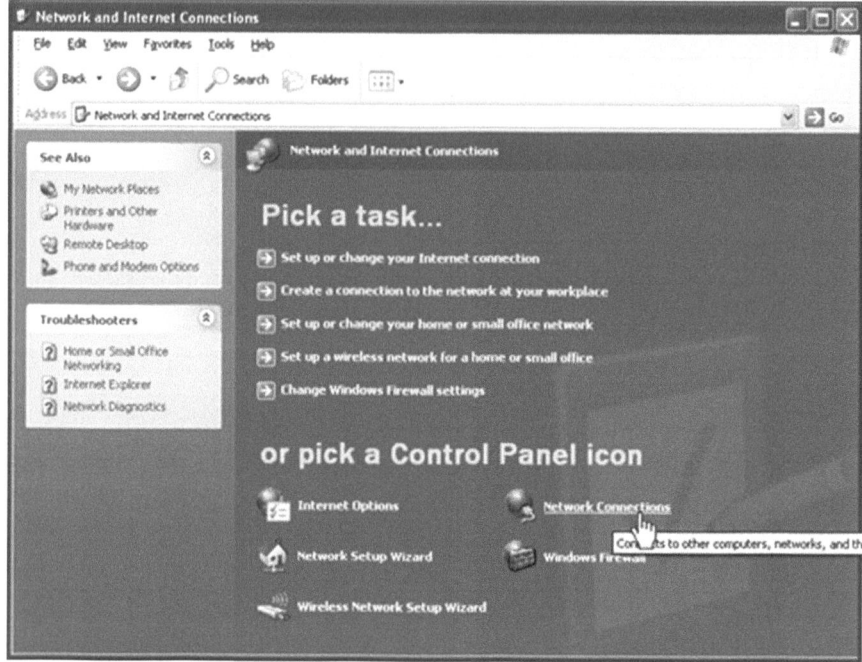

Fig. G.28 Windows XP Network and Internet Connections screen (Vixen)

Right-click on "Local Area Connection" to display and click the Properties button in the displayed menu (Fig. G.29).

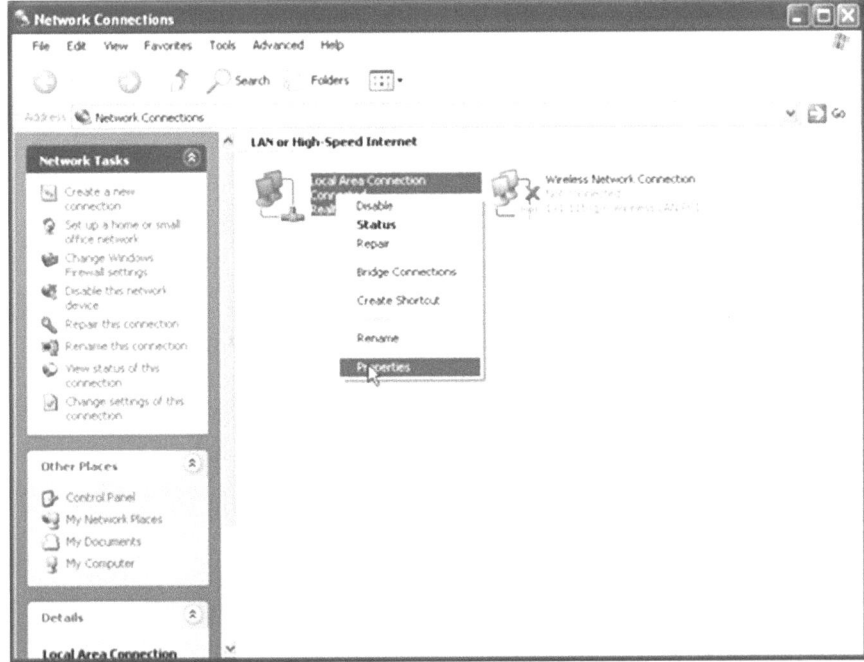

Fig. G.29 Windows XP Network connections screen (Vixen)

In the window of "This connection uses the following items", scroll down the cursor to "Internet Protocol version 4 (TCP/IPv4)" to select. Click the Properties button in the lower right of the dialog box (Figs. G.30 and G.31).

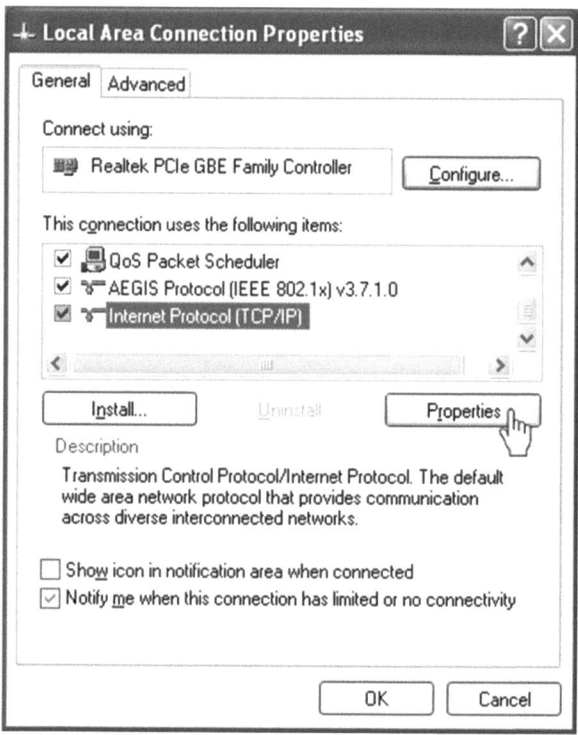

Fig. G.30 Windows XP Local Area Connection Properties screen (Vixen)

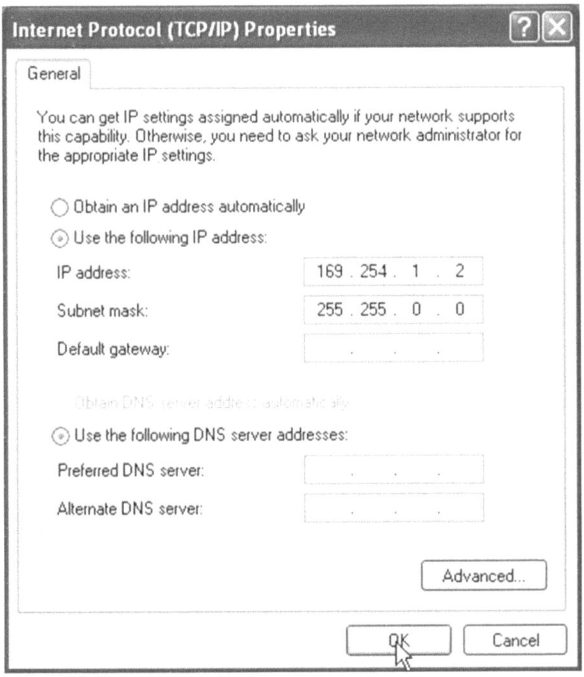

Fig. G.31 Windows XP Internet Protocol Properties screen (Vixen)

"Internet Protocol Version 4 (TCP/IPv4) Properties" is displayed. Mask check boxes for "Use the following IP Address" and enter the IP address and Subnet Mask numbers as follows:

IP Address: 169.254.1.2
Subnet Mask: 255.255.0.0

Click the OK button and the settings are completed. Go forward to 6 in the procedure on How to Install the Update Program.

When connecting to a LAN environment or a router

If connected to a LAN environment or a router that has a DHCP server, STAR BOOK will acquire the network address from DHCP server automatically at the startup.

Execute the followings when DHCP server exists.

If DHCP server does not exist, execute the procedure for a one-to-one connection.

(If the network environment is unknown, contact the administrator or the manufacturer of the router.)

Connect STAR BOOK to a LAN environment or a router.

Turn on the power of STAR BOOK.

Go to the procedure 6 if using one-to-one connection.

※ Specification and appearance may change without a notice for improvement.

© 1997–2013 VIXEN CO., Ltd All Rights Reserved.

Appendix H

Entering Comet
Orbital Elements
in the Original
Star Book

The following are the procedures for updating the original Star Book with the orbital elements of comets. These procedures are provided by Vixen at:

http://www.vixen.co.jp/en/at/update/sb-comet.htm

Note: The author is not responsible for the accuracy of these procedures. These are provided as a courtesy, and it is recommended that the user consult the Vixen dealer and review these procedures on-line prior to attempting any update. If the user's computer competency is in doubt, please do not attempt and refer this to the Vixen dealer. The photos of the screen were also taken from the website, and may not be readable in this book. Again, the author stresses before attempting any update, please refer to the Vixen website for the most up-to-date procedures and the most clear pictures.

Things to Do Before You Register the Orbital Elements of Comets

Update your STAR BOOK to the latest program version.

The latest program version is downloadable from the following site.

© Springer International Publishing Switzerland 2016

239

J.L. Chen, A. Chen, *The Vixen Star Book User Guide*, The Patrick Moore Practical Astronomy Series, DOI 10.1007/978-3-319-21593-8

Preparation for Entering Orbital Elements of Comets

Requirements

STAR BOOK controller
  Power supply for STAR BOOK
  PC (Computer) which is capable of LAN connection
  LAN crossover cable
  File of orbital elements of comets

System Requirements for PC

OS:	Microsoft Windows®XP Home Edition, XP Professional, Vista, 7 Home Edition, 7 Professional, 7 Ultimate, 8, 8 Pro
WebBrowser:	Refer to System requirements recommended by Microsoft for all version of windows OS
CPU:	Refer to System requirements recommended by Microsoft for all version of windows OS
RAM:	Refer to System requirements recommended by Microsoft for all version of windows OS
LAN:	10BASE-T/100BASE-T

※ Depending on the OS used, necessary information may not be displayed properly. If this is the case, update your OS to the latest version using Windows Update.

  ※ Microsoft and Windows are registered trademarks of Microsoft Corporation in the US and some countries.

  All product and company names mentioned in the instructions are trademarks or registered trademarks of the respective companies.

How to Install the Comet Registration File

- Download in advance the orbital elements file to the PC from here (Fig. H.1).

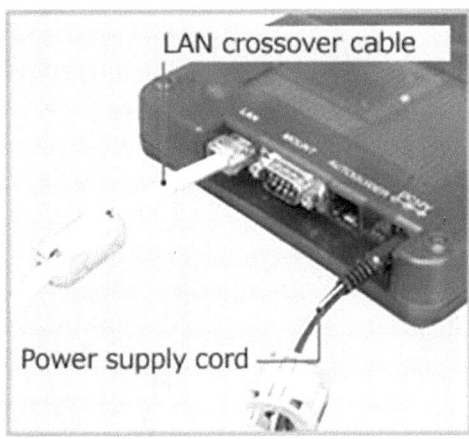

Fig. H.1 Star Book cable connections (Vixen)

- Directly connect the STAR BOOK to the PC with a LAN crossover cable.
- ※Do not power on the STAR BOOK and PC at this early stage.
- ※Regarding the connection of the LAN crossover cable to the PC, refer to the instructions provided with your PC.
- Power on the STAR BOOK. Connecting the power supply cord will start up the STAR BOOK. Turn on the PC.
- Wait for a period of 1 min as the STAR BOOK and PC will automatically acquire information on networking.

• Advance the dialogs on screen of the STAR BOOK until the star chart appears in Chart Mode or Scope Mode (Figs. H.2 and H.3).

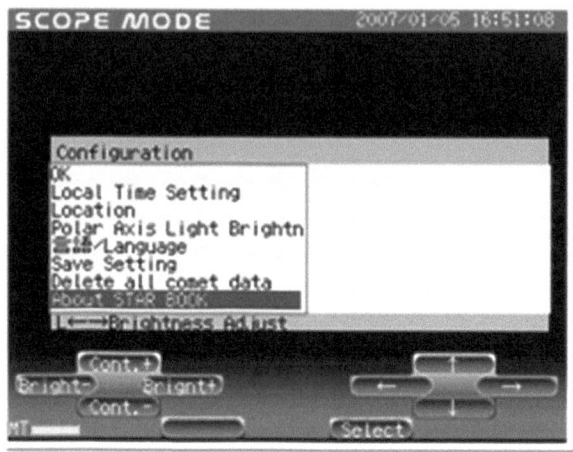

Fig. H.2 Star Book Configuration (Vixen)

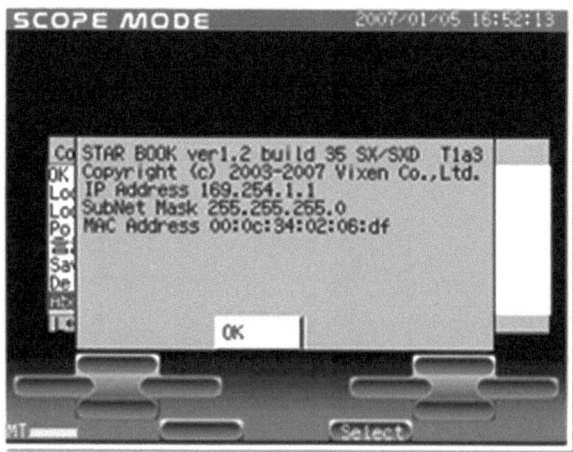

Fig. H.3 Star Book Version screen (Vixen)

- In the System Menu of the STAR BOOK, scroll down the cursor to select "About STAR BOOK" and press Select key to enter.

  - IP Address: 169.254.a.b (a, b are arbitrary numbers)
  - SubNet Mask: 255.255.0.0 (Figs. H.4 and H.5)

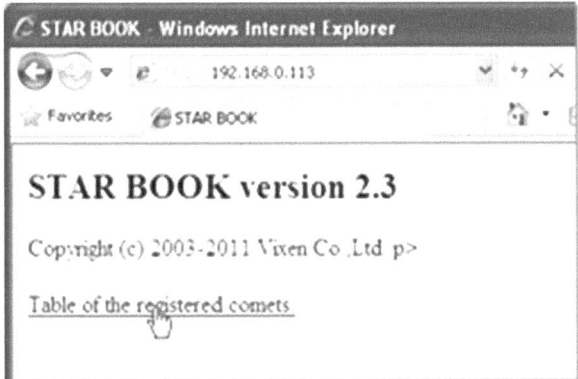

Fig. H.4 Internet Explorer page (Vixen)

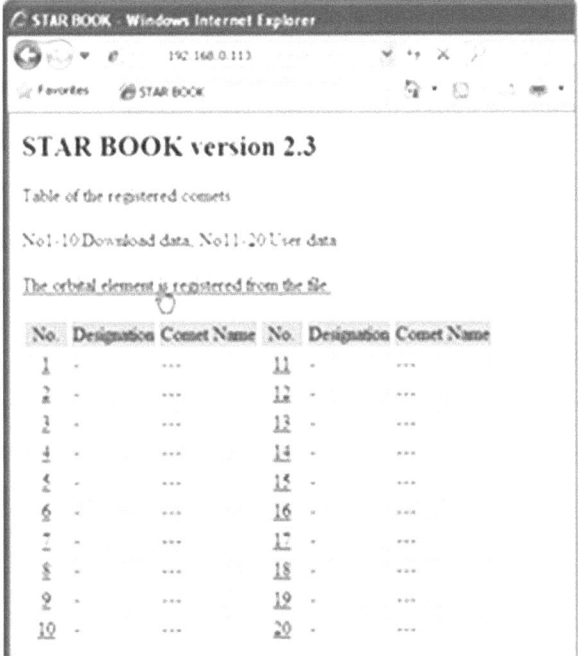

Fig. H.5 Table of registered comets (Vixen)

- Open Internet Explorer and enter the IP address you obtained in step 4 into the address bar to display the entry page.
- Example: If the IP address is 192.168.0.189, put http://192.168.0.189 on the address bar.

  ※Change the setting to have the address bar appear if it is hidden on your PC.

- The entry page shown on the left appears. Click "Table of the registered comets" in the entry page (Fig. H.6).

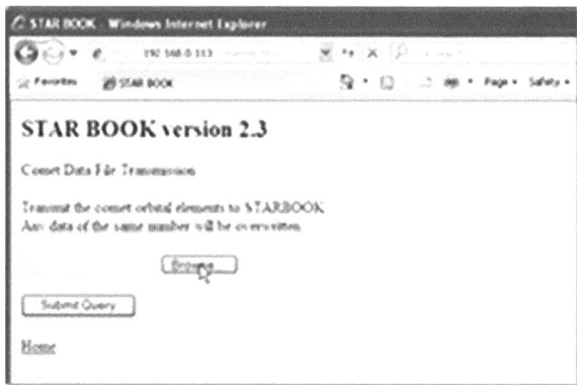

Fig. H.6 Data File Transmission from the File (Vixen)

- The "Table of the registered comets" dialog box appears. Confirm that the orbital elements of the comets are registered in the data track in order starting from the number 1 on the table. The existing data are always overwritten by the new information. You need to transfer the existing data to an available track so it is not lost.
- Click "Data File Transmission from the File" to access the entry dialog box.

• The "Data File Transmission from the File" page appears. Click on the Browse…
  button to continue (Figs. H.7 and H.8).

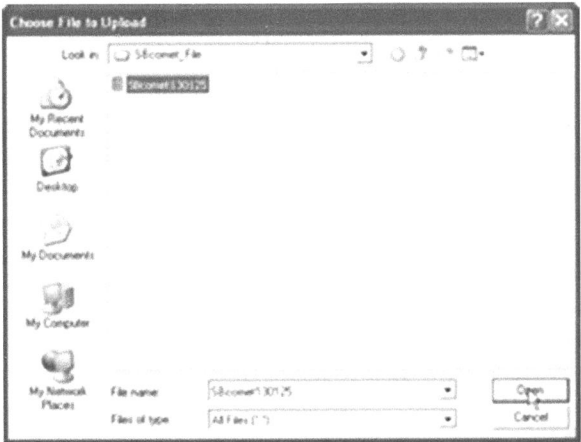

Fig. H.7 Open file (.txt) of the orbital elements of comets (Vixen)

Fig. H.8 Star Book Version screen (Vixen)

- Designate the folder that contains a file (.txt) of the orbital elements of comets. Point to the file of the orbital elements of comets and click the Open button.
- Example: SBcomet2009-01.txt
- Click on the Submit Query button to send.
- You will see the message below at the end of a successful transmission to your STAR BOOK (Figs. H.9 and H.10).

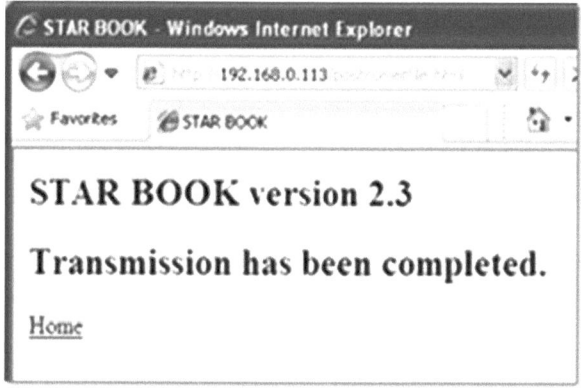

Fig. H.9 Transmission has been completed (Vixen)

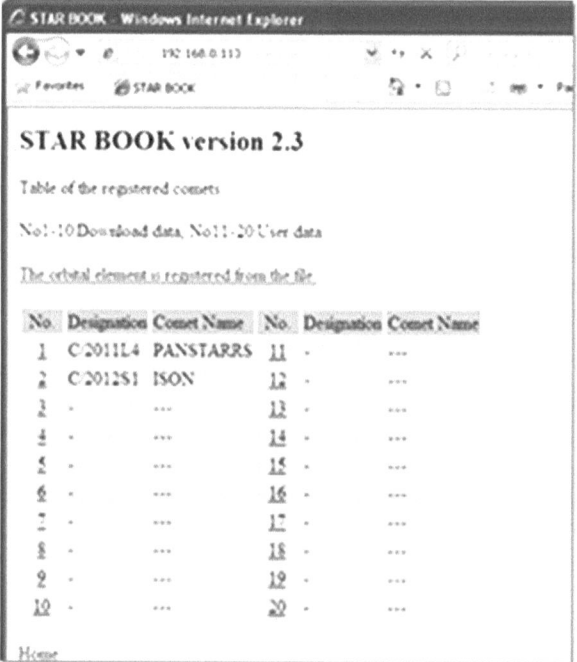

Fig. H.10 Display the table of the registered comets (Vixen)

- Display the table of the registered comets to confirm the new registration (Fig. H.11).

Fig. H.11 Orbital Element details (Vixen)

- Clicking the number in the table will show you the orbital elements of a comet in details.
- Be sure to reboot the STAR BOOK after you finish registering the orbital elements of comets.

In Case of Your Download Is Unsuccessful

Follow the directions below to change setting on your PC (This requires an authorization from administrator.) (Figs. H.12 and H.13)

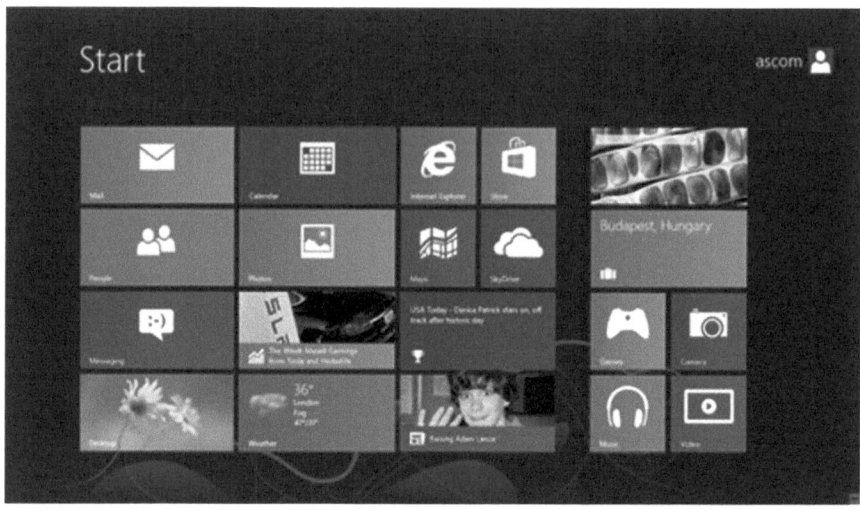

Fig. H.12 Windows 8 Start screen (Vixen)

Fig. H.13 Windows 8 Start Screen cont'd (Vixen)

How to Conform Your IP Address (for Windows 8)

Right-click on the background in the start up screen. Click on All apps icon at the bottom right of the screen to display every application (Figs. H.14 and H.15).

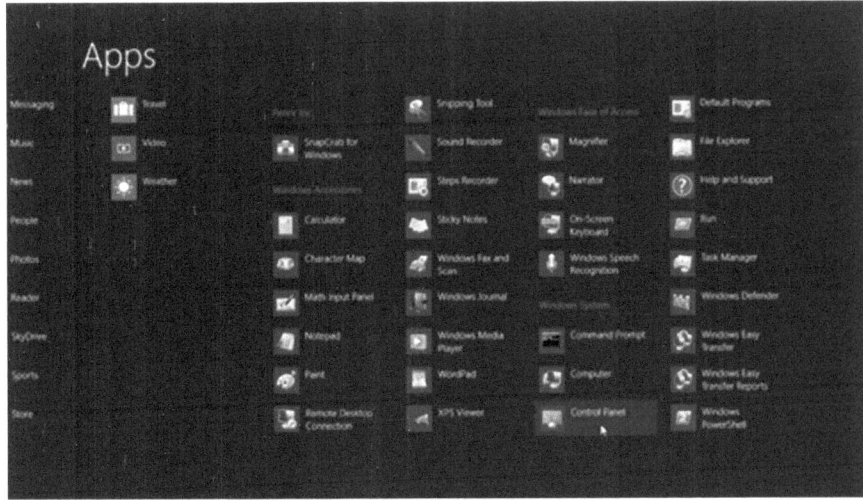

Fig. H.14 Windows 8 Apps screen (Vixen)

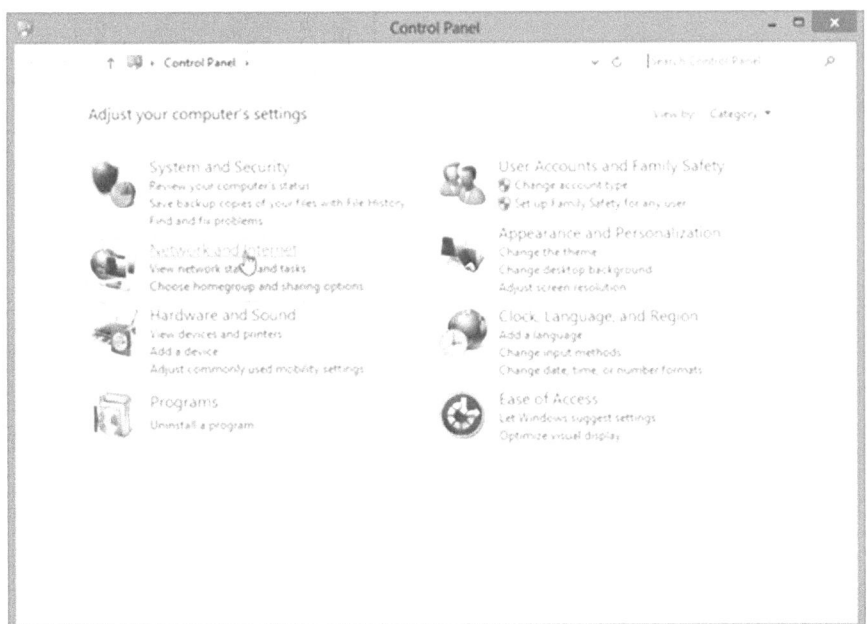

Fig. H.15 Windows 8 Control Panel (Vixen)

Select "Control Panel" and move to "Network and Internet" (Fig. H.16).

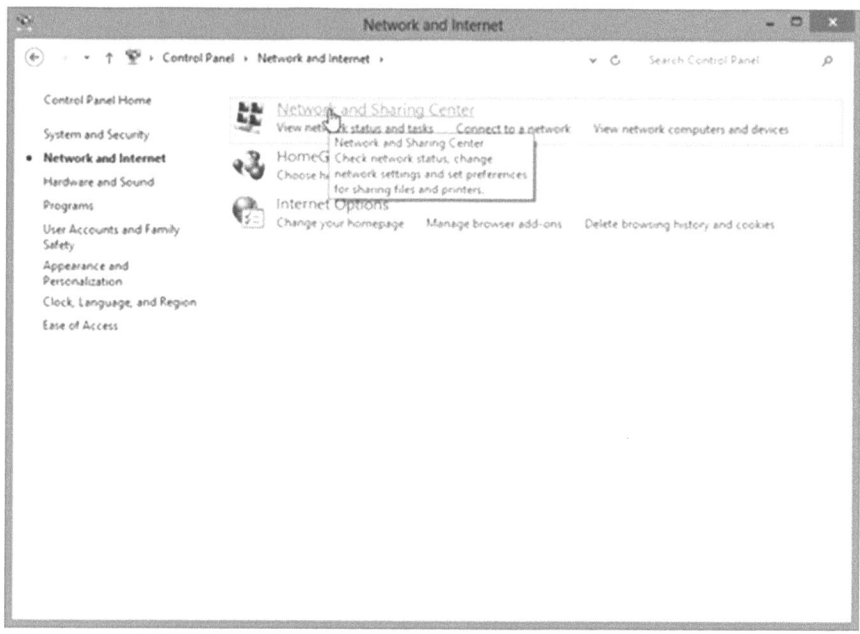

Fig. H.16 Windows 8 Network and Internet (Vixen)

Click on "Network and Sharing Center". The screen of "View your basic network information and set up connections" is displayed (Figs. H.17, H.18, and H.19).

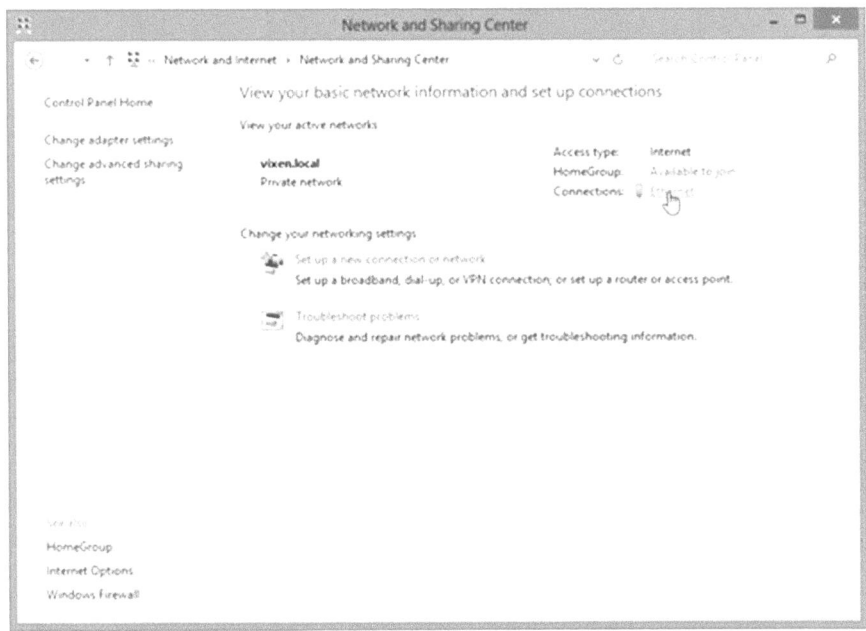

Fig. H.17 Windows 8 Network and Sharing Center (Vixen)

Fig. H.18 Windows 8 Ethernet Status (Vixen)

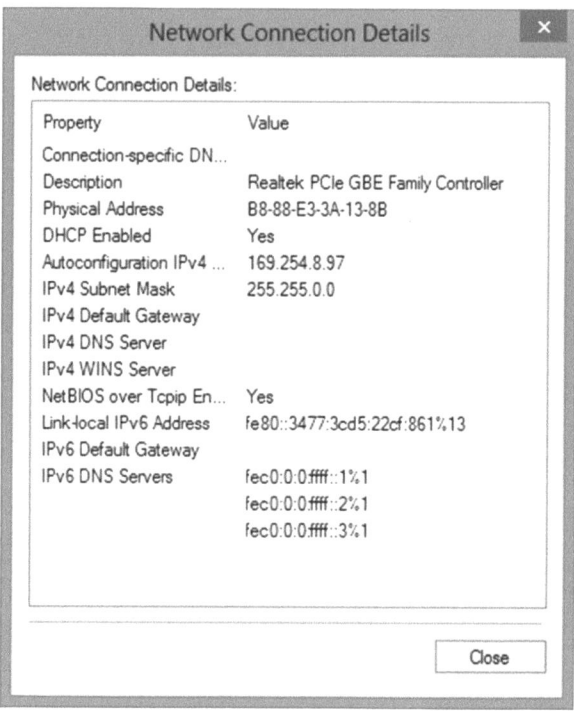

Fig. H.19 Windows 8 Network Connection Details screen (Vixen)

Display "Local Area Connection—Details (E)" and confirm the following settings.

- IPv4 IP Address: 169.254.a.b (a, b are arbitrary numbers)
- IPv4 Subnet Mask: 255.255.0.0

If different numbers are set on your PC, or if the DHCP server is not available as it is selected to "NO", there is a possibility that a check mark is removed from "Obtain the IP address automatically". If this is the case, follow the procedure below to change the settings and obtain the IP address automatically (Fig. H.20).

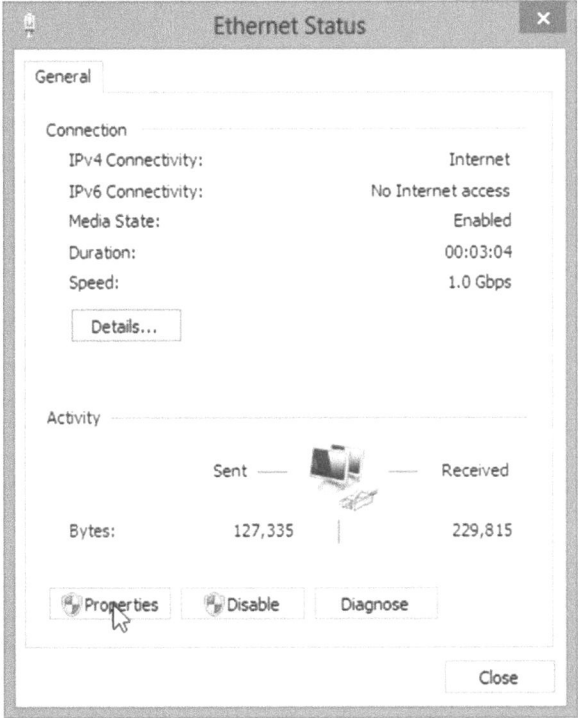

Fig. H.20 Windows 8 Ethernet Status (Vixen)

Click the properties button on "Local Area Connection Status" dialog box to open (Fig. H.21).

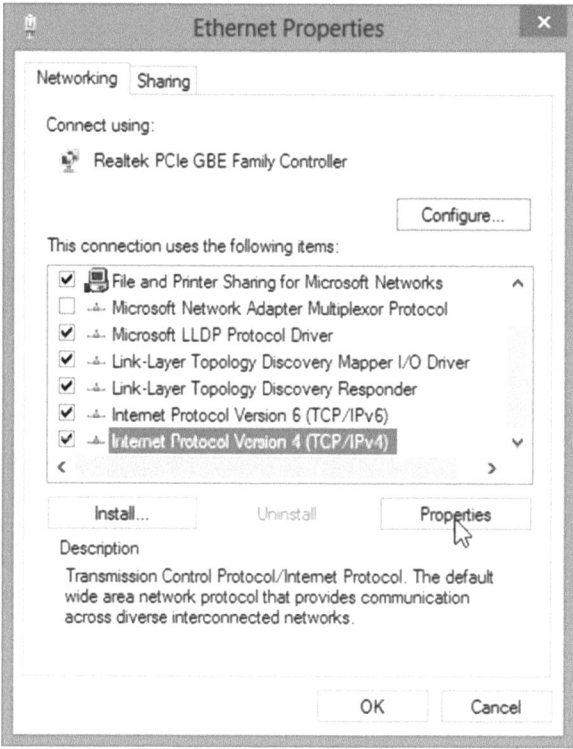

Fig. H.21 Windows 8 Etherenet Properties (Vixen)

Scroll down the cursor to "Internet Protocol version 4 (TCP/IPv4)" to select. Click the properties (Fig. H.22).

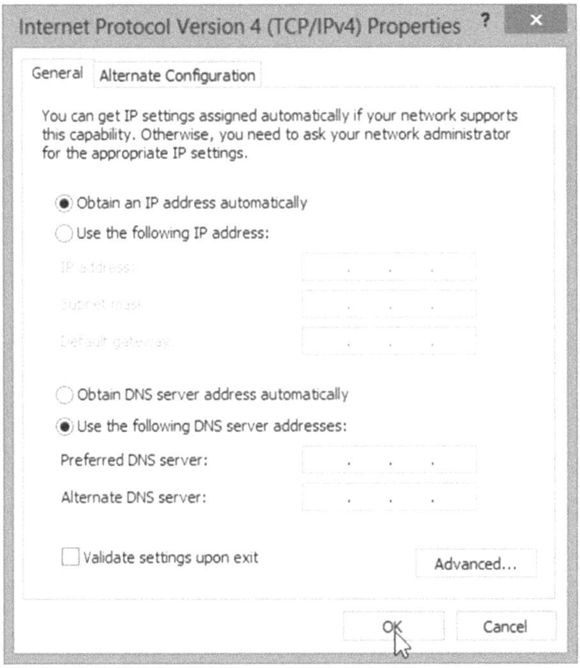

Fig. H.22 Windows 8 Internet Protocol Properties (Vixen)

Mark check boxes for "Obtain an IP address automatically" and "Obtain DNS server address automatically" in "General" tab. Click the OK button (Fig. H.23).

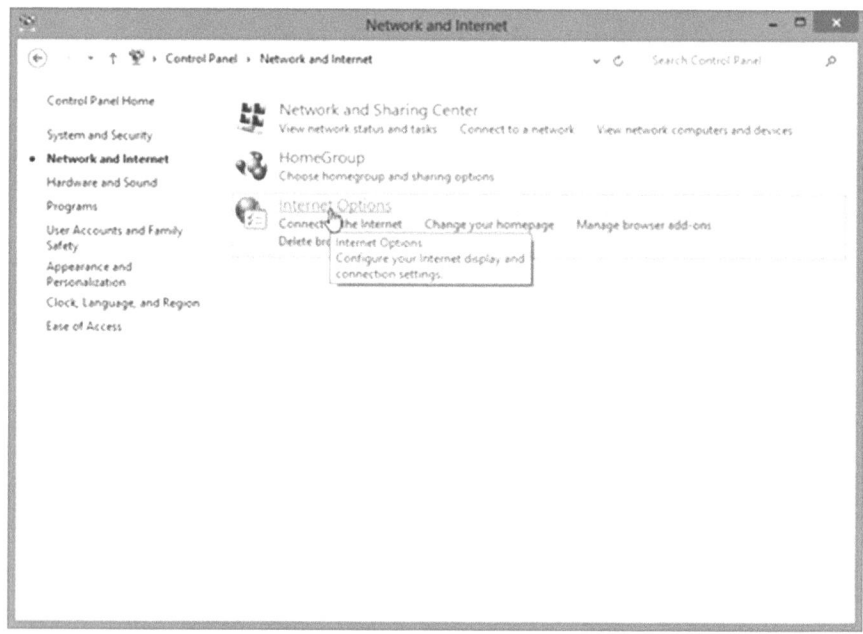

Fig. H.23 Windows 8 Network and Internet (Vixen)

Go to "Control Panel" and select "Internet Option" in "Network and Internet" dialog box (Fig. H.24).

Fig. H.24 Windows 8 Internet Options (Vixen)

"Internet Protocol Properties" are displayed. Move to "Connection" tab and select "LAN Settings".

Confirm that a check box for "Use Proxy Server for LAN" in the lower center is not market. If the box is checked, remove the check mark. (Remember to return to the previous setting when you finish the download.)

Go back to the procedure on how to install the Comet Registration File.

Click "Start" to open the window and select "Control Panel (C)" (Figs. H.25 and H.26)

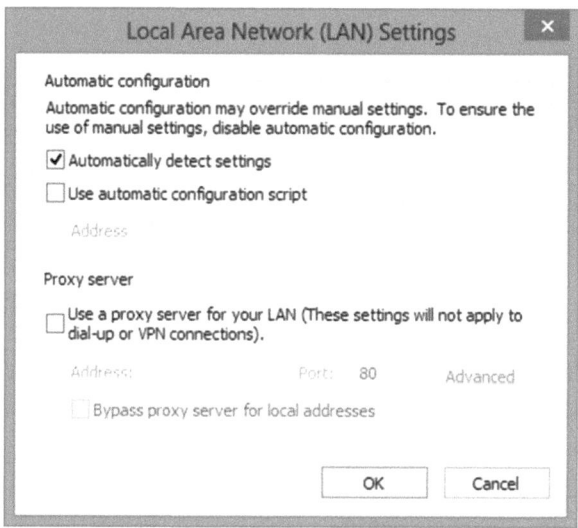

Fig. H.25 Windows 7 LAN Settings (Vixen)

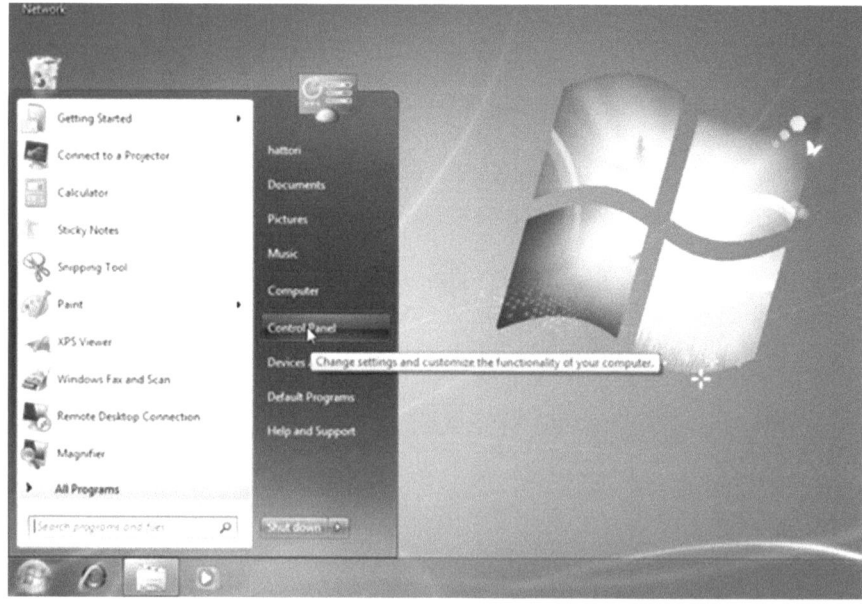

Fig. H.26 Windows 7 Start screen (Vixen)

Select "Network and Internet" (Fig. H.27).

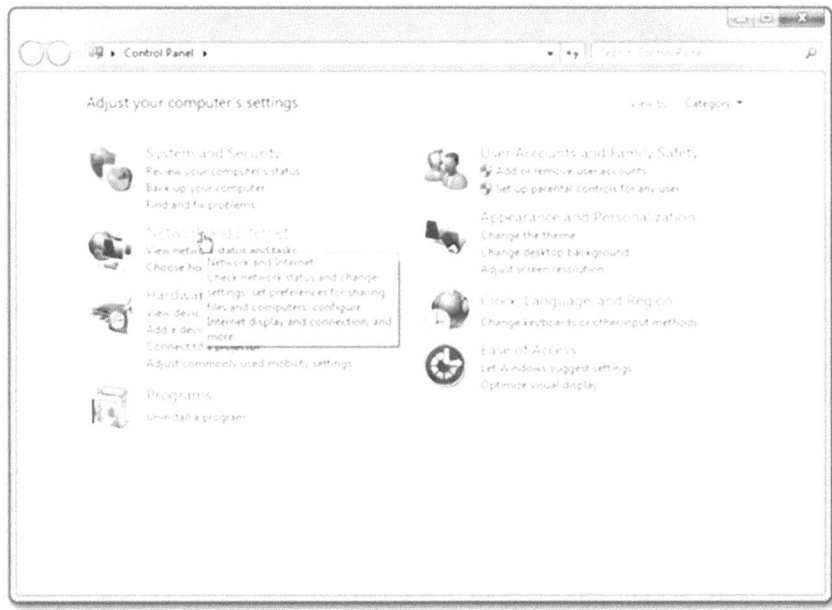

Fig. H.27 Windows 7 Control Panel (Vixen)

Click on "Network and Sharing Center" (Figs. H.28, H.29, and H.30).

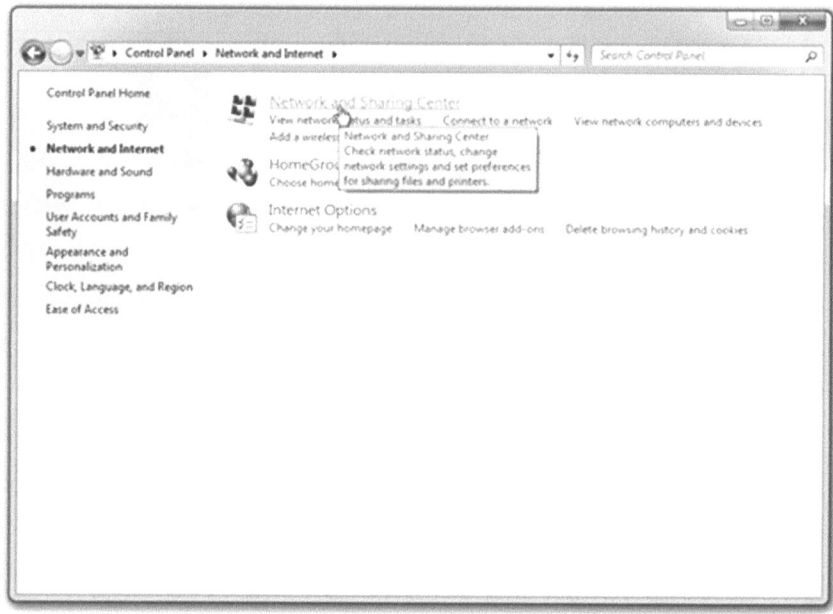

Fig. H.28 Windows 7 Network and Internet screen (Vixen)

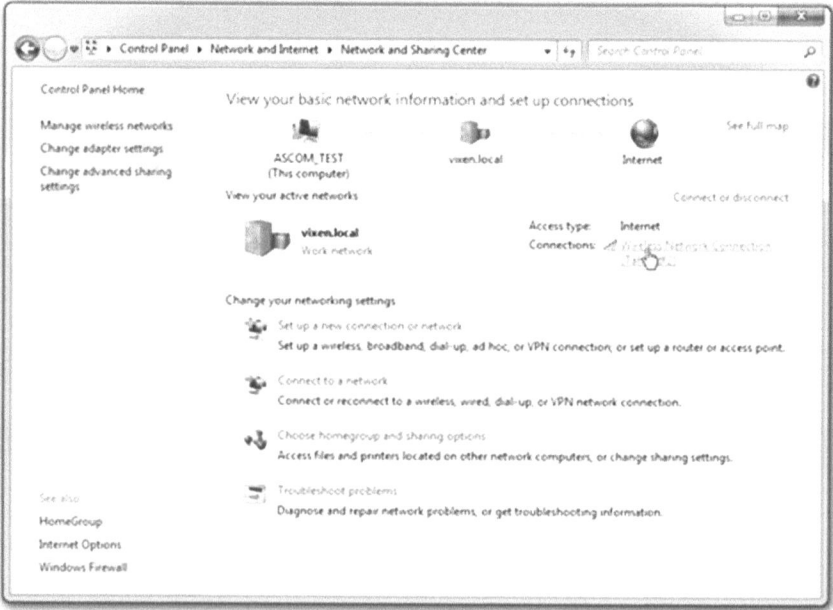

Fig. H.29 Windows 7 Network and Sharing Center (Vixen)

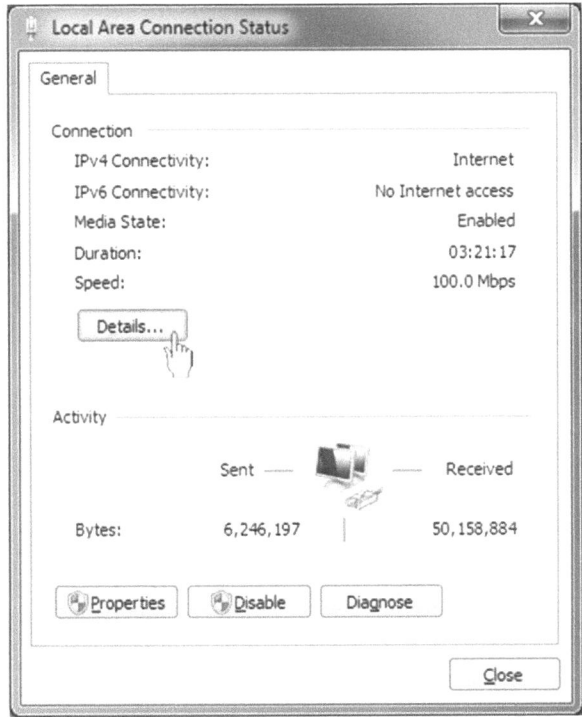

Fig. H.30 Windows 7 Local Area Connection Status (Vixen)

Go to "Local Area Connection—Details (E)".
Confirm that the following numbers are set in the dialog box.

- IPv4 IP Address: 169.254.a.b (a, b are arbitrary numbers)
- IPv4 Subnet Mask: 255.255.0.0

If different numbers are set on your PC, or if the DHCP server is available as it is selected to "NO", there is a possibility that a check mark is removed from "Obtain IP address automatically". If this is the case, follow the procedure below to change the settings and obtain the IP address automatically (Fig. H.31).

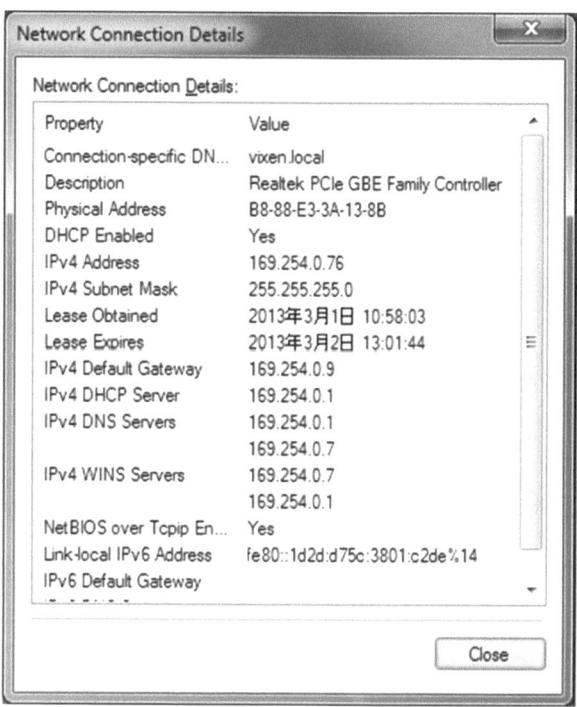

Fig. H.31 Windows 7 Network Connection Details screen (Vixen)

Click the Properties button in the "Local Area Connection Status" dialog box (Fig. H.32).

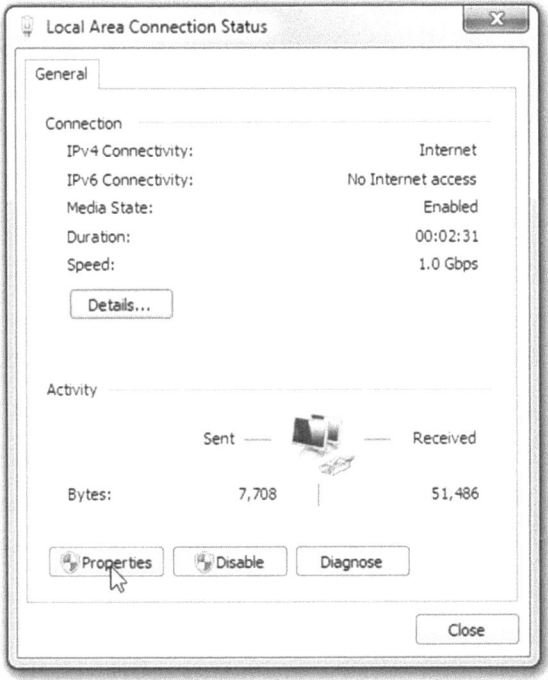

Fig. H.32 Windows 7 Local Area Connection Status (Vixen)

In the window of "This connection uses the following items (O)", select "Internet Protocol version 4 (TCP/IPv4)". Click the Properties button (Fig. H.33).

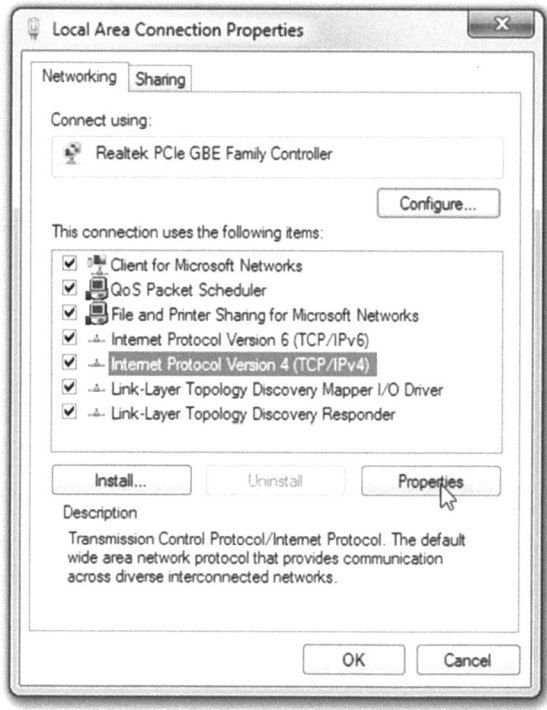

Fig. H.33 Windows 7 Local Area Connection Properties (Vixen)

In "General tab", mark check boxes for "Obtain IP address automatically (O)" and "Obtain DNS server address automatically (B)". Click the OK button (Fig. H.34).

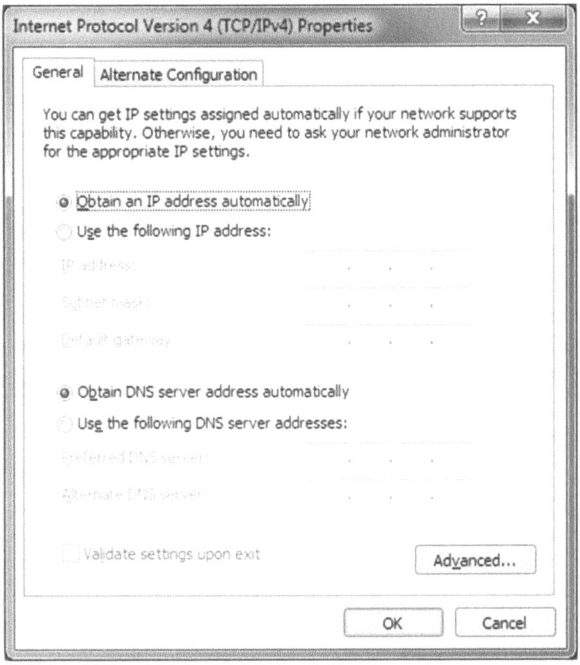

Fig. H.34 Windows 7 Internet Protocol Properties screen (Vixen)

Go to "Control Panel" and select "Internet Option" in "Network and Internet" dialog box (Fig. H.35).

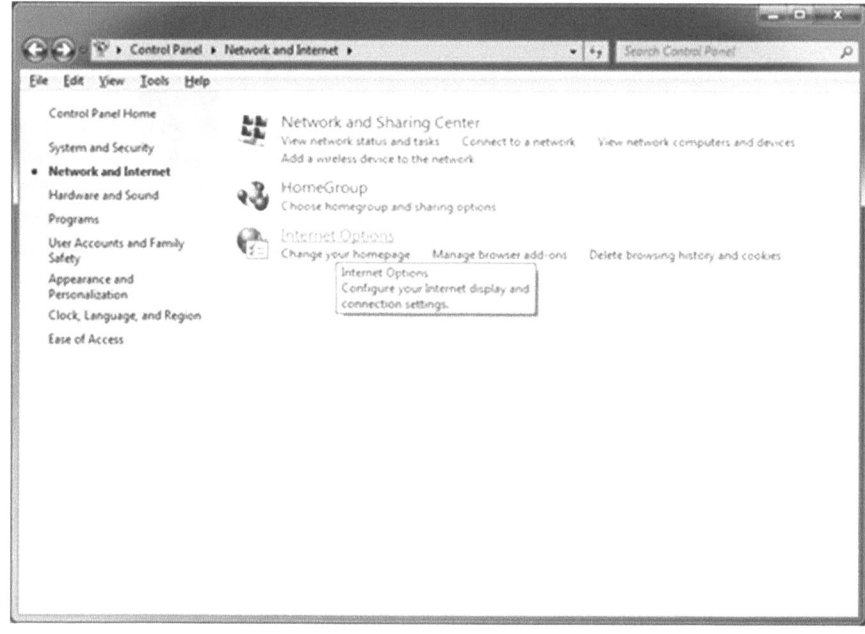

Fig. H.35 Windows 7 Network and Internet screen (Vixen)

"Internet Protocol Properties" are displayed. Go to "Connection" tab select "LAN Settings" (Fig. H.36).

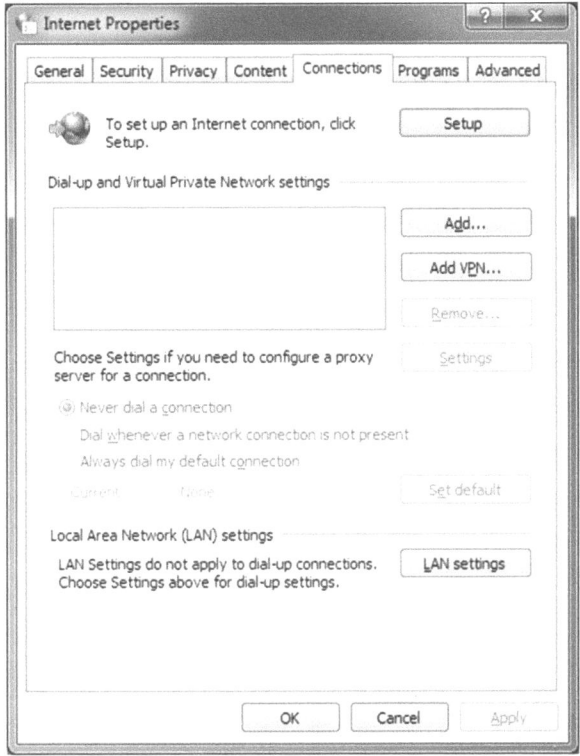

Fig. H.36 Windows 7 Internet Properties (Vixen)

Confirm that a check box for "Use Proxy Server for LAN" in the lower center is not marked. If the box is checked, remove the check mark. (Remember to return to the previous settings when you finish the download.)

Go back to the procedure on "How to Install the Comet Registration File".

Click "Start" to open the window and select "Control Panel (C)" (Figs. H.37 and H.38).
  Select "Network and Internet" (Fig. H.38).

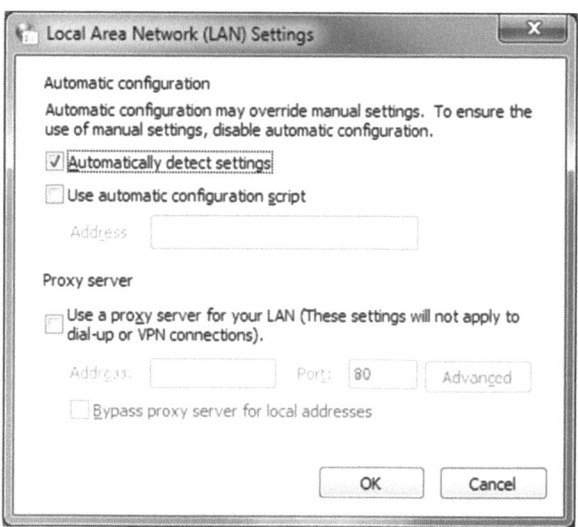

Fig. H.37 Windows Vista (Vixen)

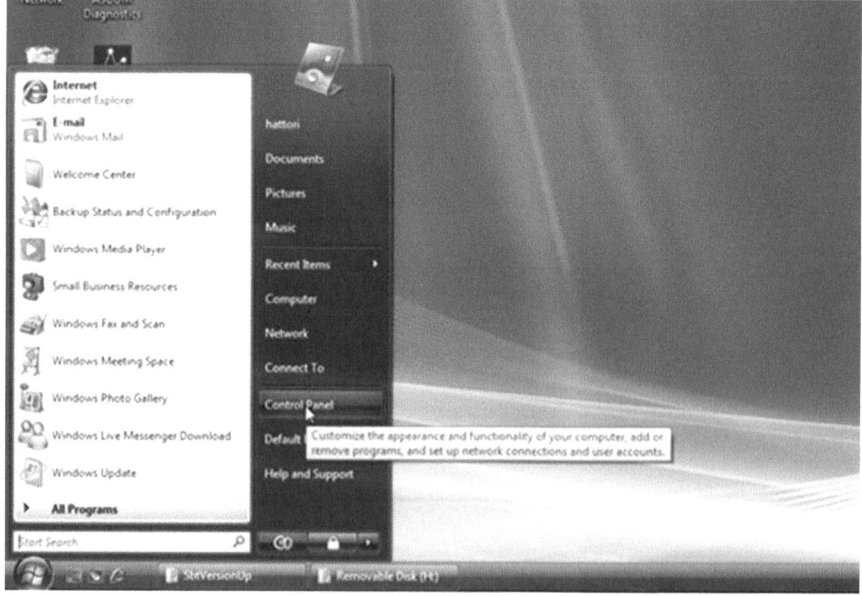

Fig. H.38 Windows Vista Start screen (Vixen)

Click on "Network and Sharing Center" (Fig. H.39).

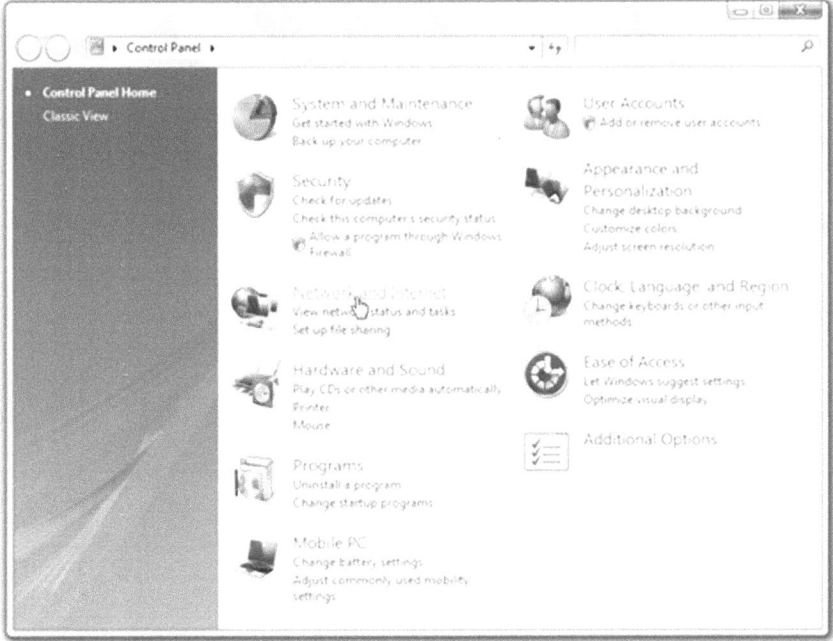

Fig. H.39 Windows Vista Control Panel (Vixen)

Click on "Manage Network Connections" in the right column (Fig. H.40, H.41, and H.42).

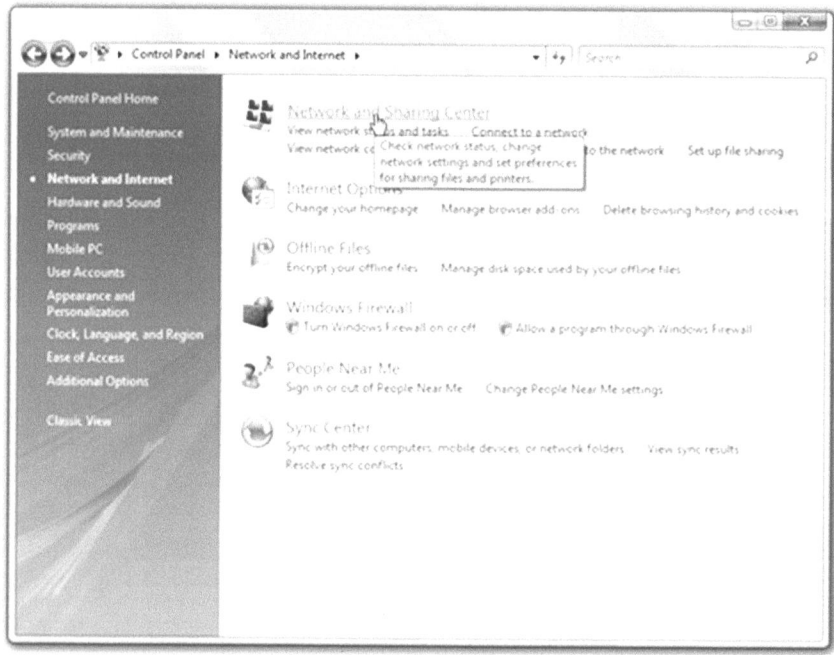

Fig. H.40 Windows Vista Network and Internet (Vixen)

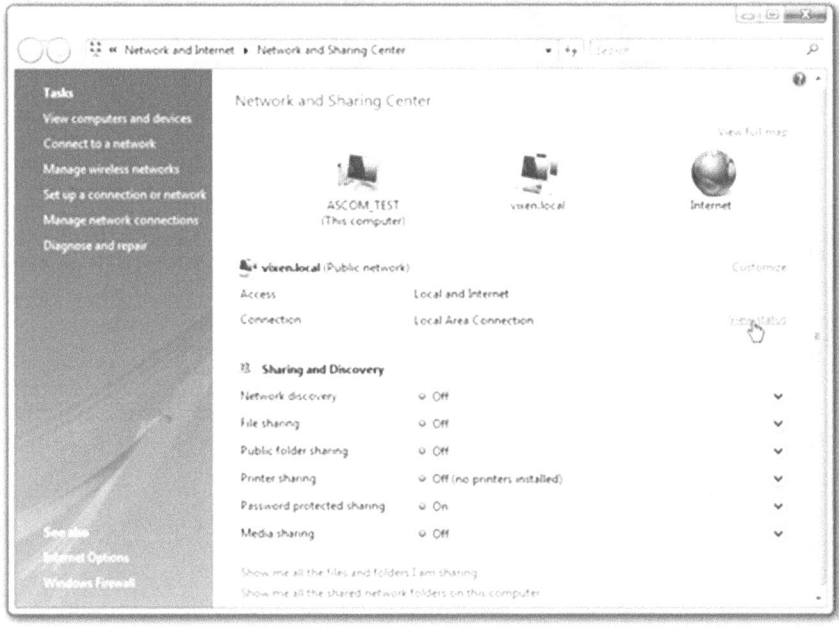

Fig. H.41 Windows Vista Network and Sharing Center (Vixen)

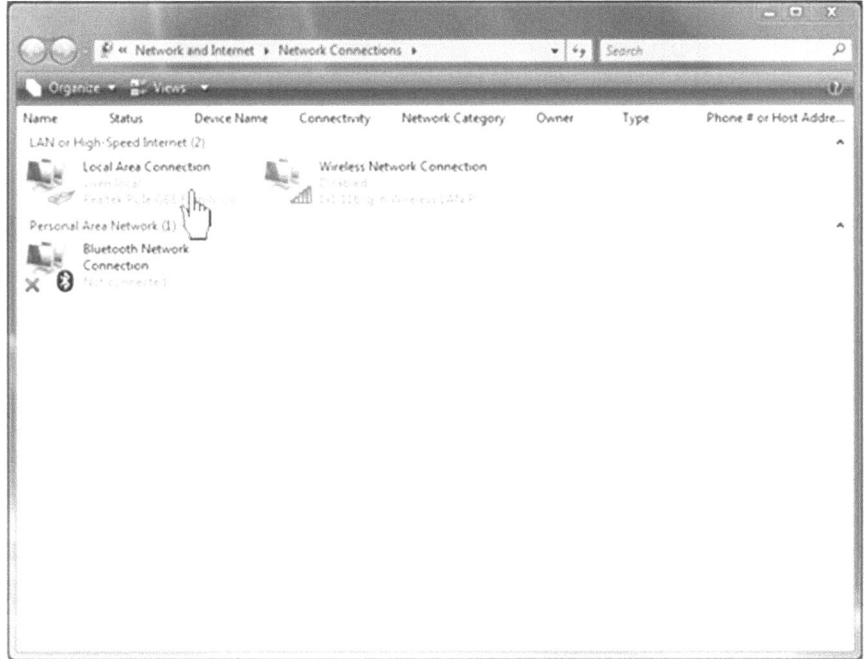

Fig. H.42 Windows Vista Network Connections (Vixen)

Double-click on "Local Area Connection" and open the "Local Area Connection Status" dialog box.

Click the Details (E) and confirm that the following numbers are set in the dialog box.

- IPv4 IP Address: 169.254.a.b (a, b are arbitrary numbers)
- IPv4 Subnet Mask: 255.255.0.0

If different numbers are set on your PC, or if the DHCP server is available as it is selected to "NO", there is a possibility that a check mark is removed from "Obtain IP address automatically". If this is the case, follow the procedure below to change the settings and obtain the IP address automatically (Fig. H.43).

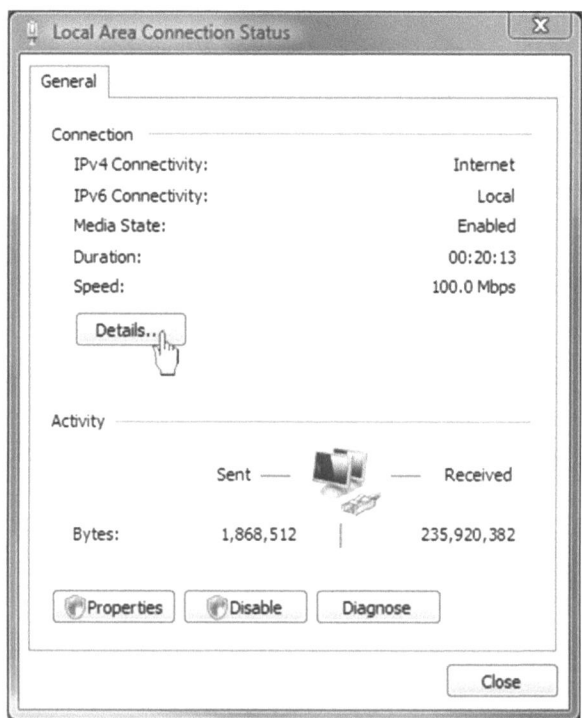

Fig. H.43 Windows Vista Local Area Connection Status (Vixen)

Click the Properties button in the "Local Area Connection Status" dialog box (Fig. H.44).

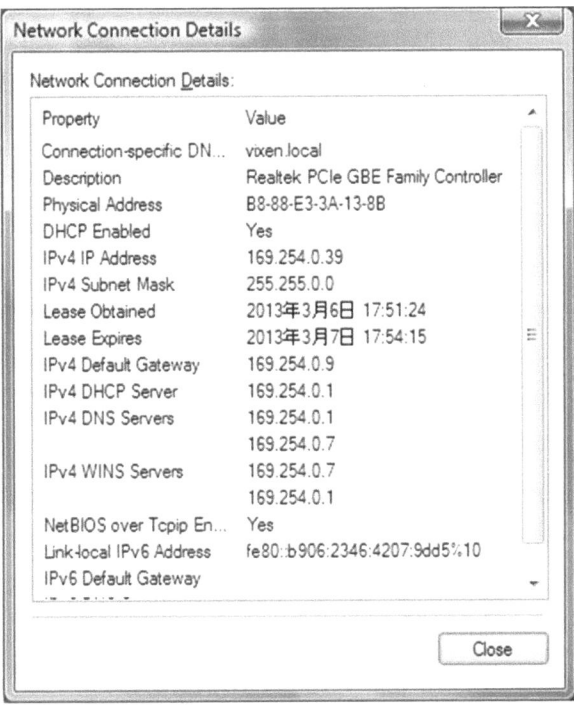

Fig. H.44 Windows Vista Network Connection Details (Vixen)

In the window of "This connection uses the following items (O)", select "Internet Protocol version 4 (TCP/IPv4)". Click the Properties button (Fig. H.45).

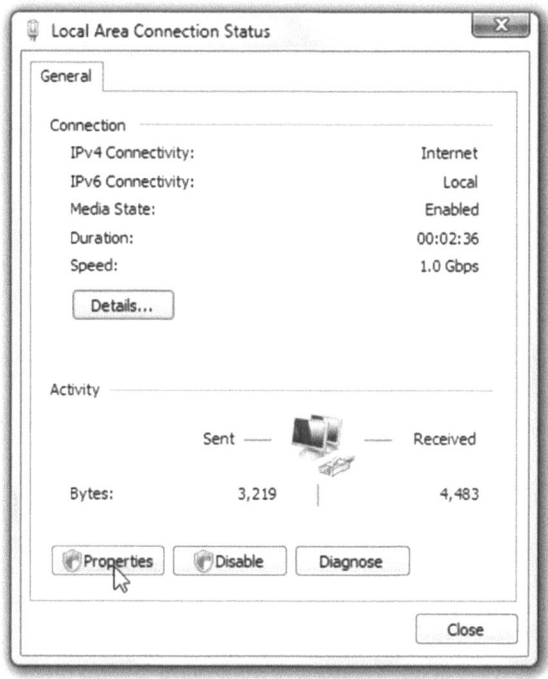

Fig. H.45 Windows Vista Local Area Connection Status (Vixen)

In "General tab", mark check boxes for "Obtain IP address automatically (O)" and "Obtain DNS server address automatically (B)". Click the OK button (Fig. H.46).

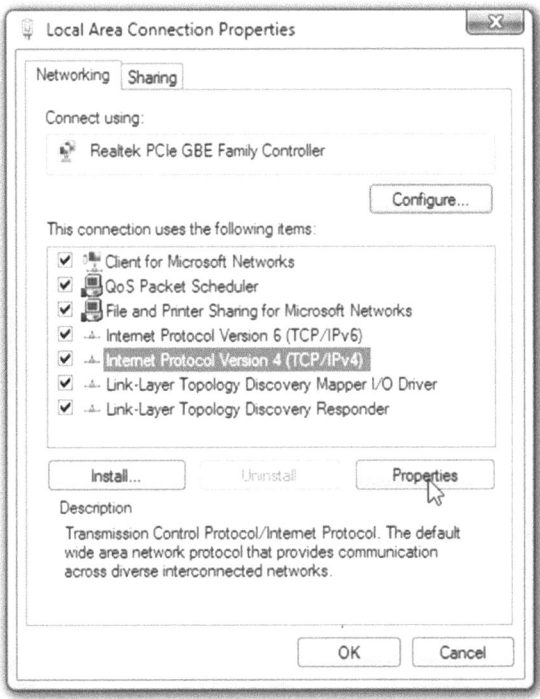

Fig. H.46 Windows Vista Local Area Connection Properties (Vixen)

Go to "Control Panel" and select "Internet Option" in "Network and Internet" dialog box (Figs. H.47 and H.48).

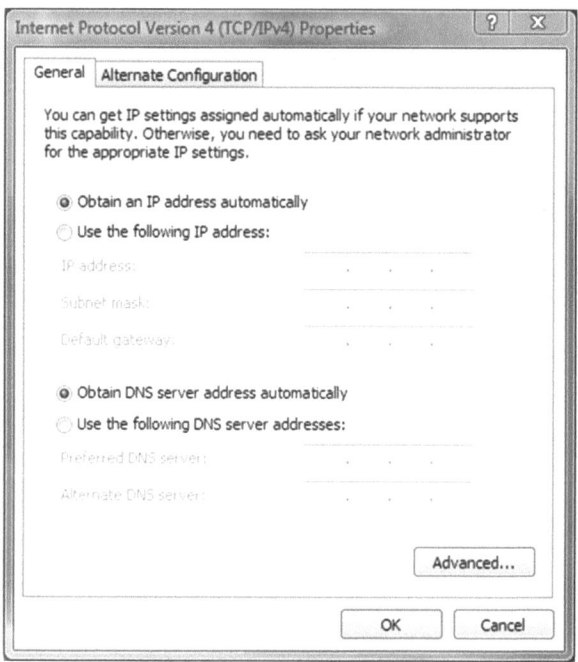

Fig. H.47 Windows Vista Internet Propocol Properties (Vixen)

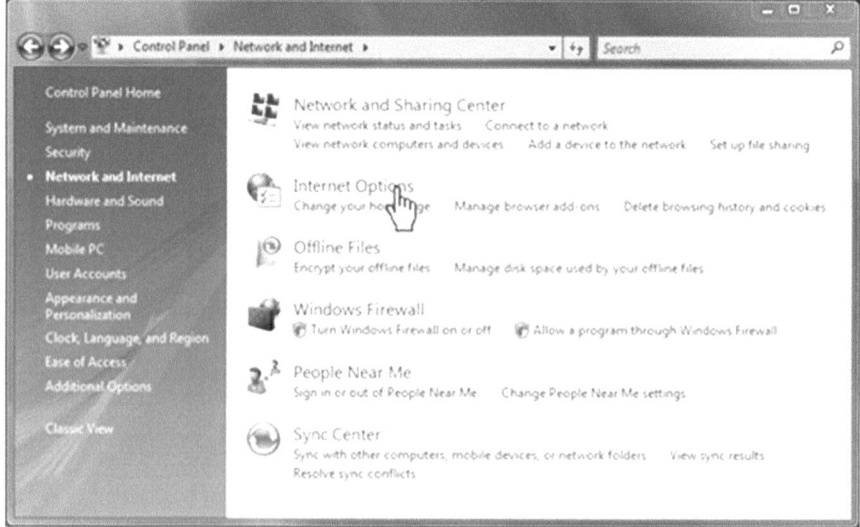

Fig. H.48 Windows Vista Network and Internet (Vixen)

"Internet Protocol Properties" are displayed. Go to "Connection" tab select "LAN Settings" (Fig. H.49).

Fig. H.49 Windows Vista Internet Properties (Vixen)

Confirm that a check box for "Use Proxy Server for LAN" in the lower center is not marked. If the box is checked, remove the check mark. (Remember to return to the previous settings when you finish the download.)

Go back to the procedure on "How to Install the Comet Registration File".

How to Conform Your IP Address (for Windows XP)

Click the Start button in the bottom left of the screen to display the Start menu. Select "Control Panel" and click on it to display (Figs. H.50 and H.51).

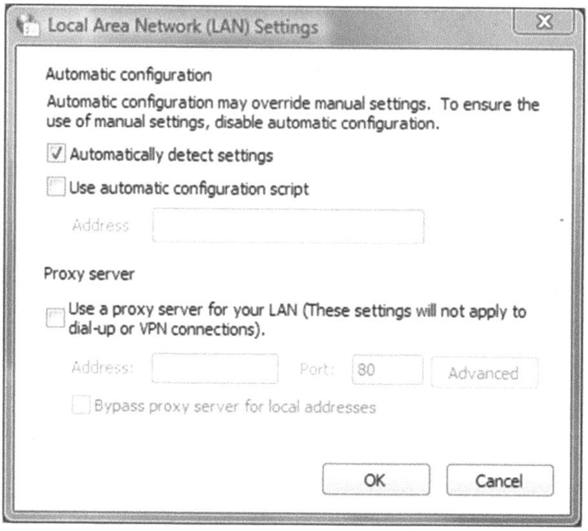

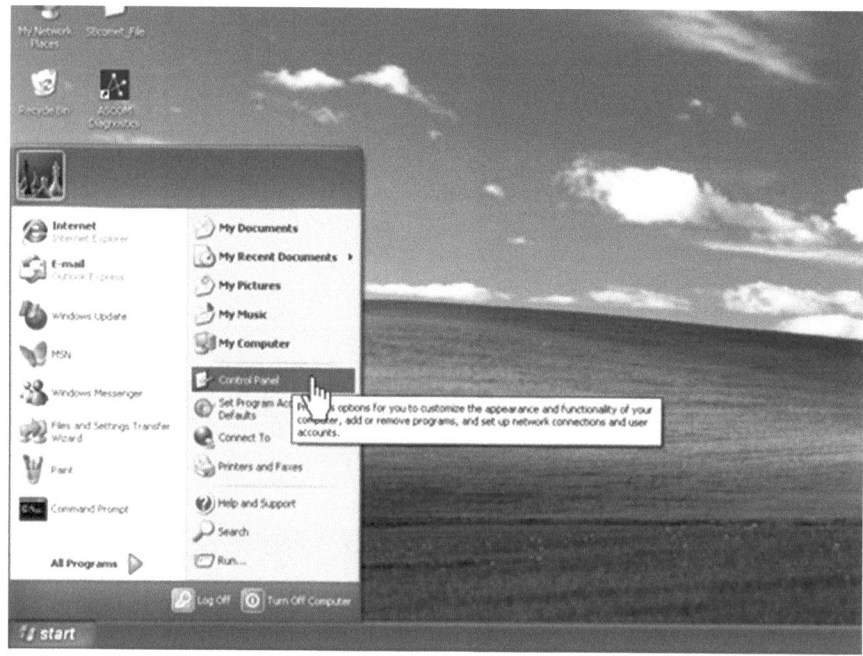

Fig. H.50 Windows XP Local Area Network settings (Vixen)

Fig. H.51 Windows XP Start screen (Vixen)

In the "Control Panel" menu, click on "Network and Internet" to display (Fig. H.52).

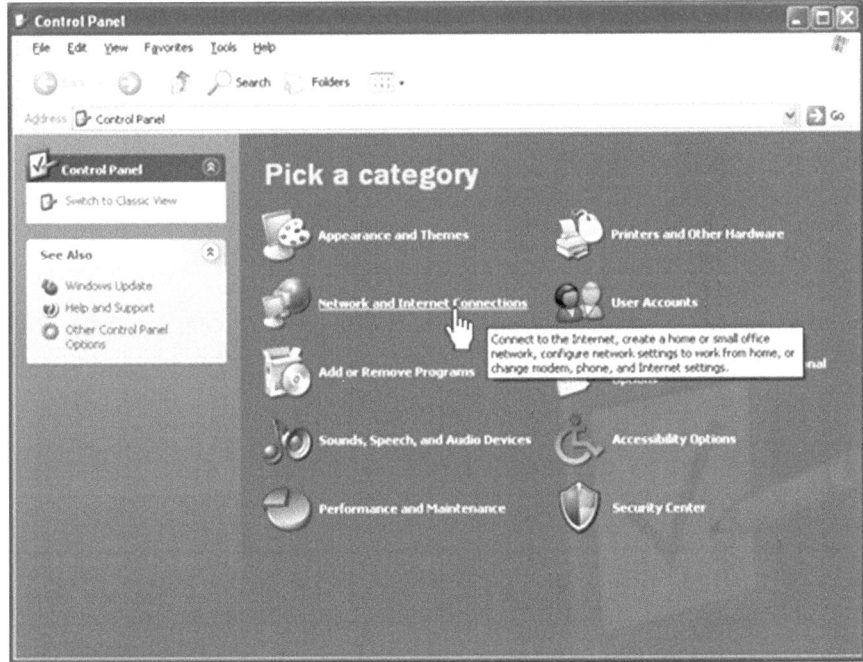

Fig. H.52 Windows XP Control Panel (Vixen)

Click on "Network Connections" (Fig. H.53).

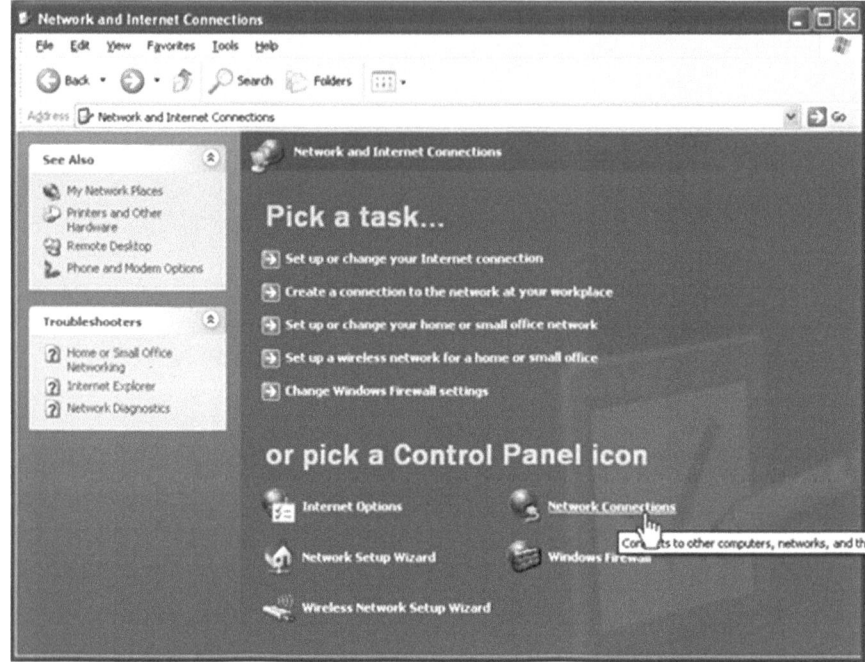

Fig. H.53 Windows XP Network and Internet Connections screen (Vixen)

Double-click on "Local Area Connection" to display (Figs. H.54 and H.55).

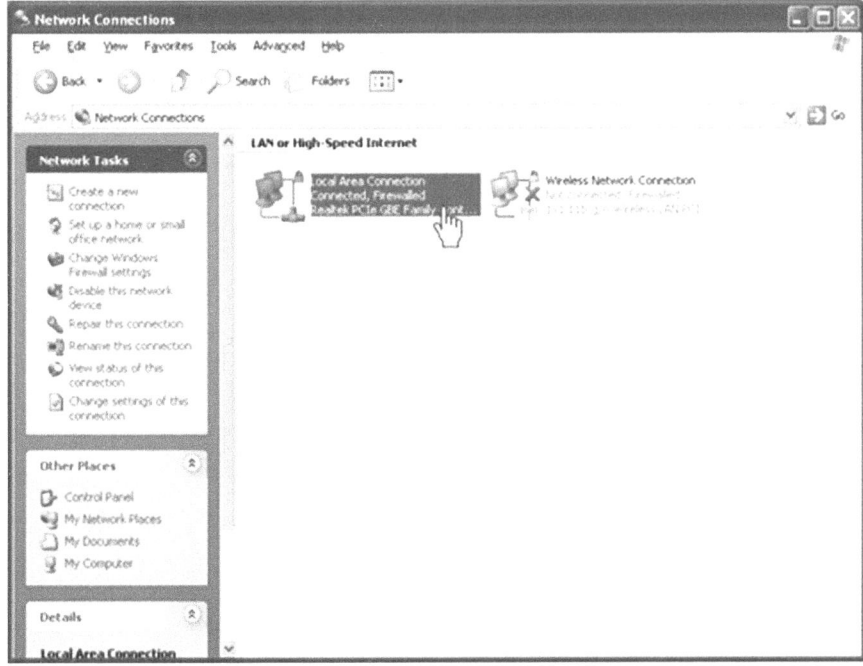

Fig. H.54 Windows XP Network Connections (Vixen)

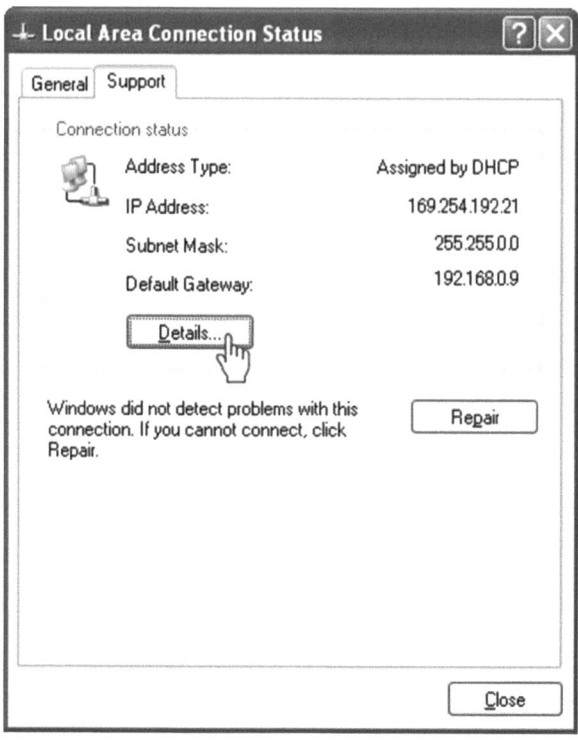

Fig. H.55 Windows XP Local Area Connection Status (Vixen)

Click the Details (E) in the Support tube in "Local Area Connection Status", and confirm the following numbers are set in the dialog box.

- IP Address: 169.254.a.b (a, b are arbitrary numbers)
- Subnet Mask: 255.255.0.0

If different numbers are set on your PC, or if the DHCP server is available as it is selected to "NO", there is a possibility that a check mark is removed from "Obtain IP address automatically". If this is the case, follow the procedure below to change the setting and obtain the IP address automatically (Fig. H.56).

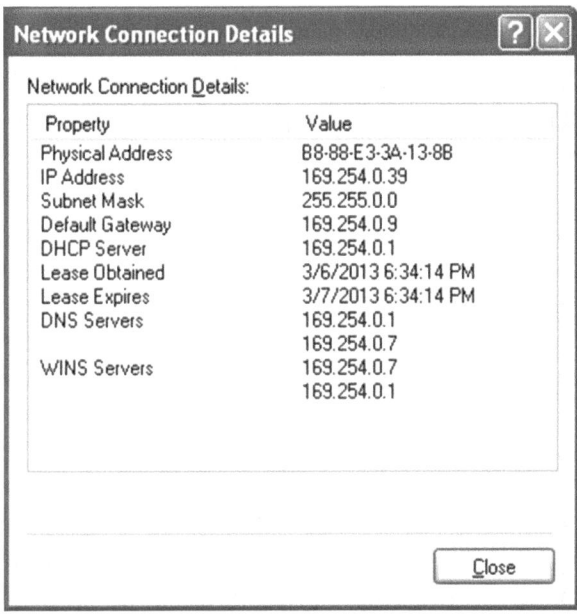

Fig. H.56 Windows XP Network Connection Details (Vixen)

Click the General tube in the "Local Area Connection Status" dialog box and click the Properties button (Fig. H.57).

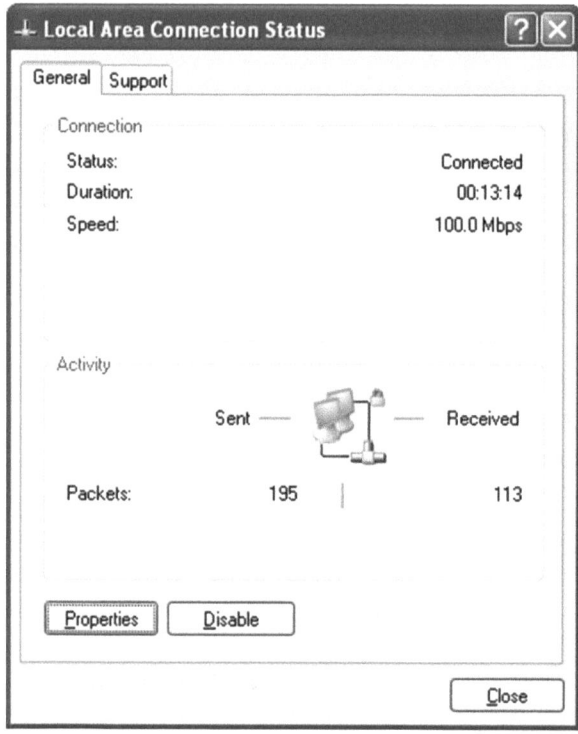

Fig. H.57 Windows XP Local Area Connection Status (Vixen)

In the windows of "This connection uses the following items (O)", select "Internet Protocol (TCP/IP)". Click the Properties button (Fig. H.58).

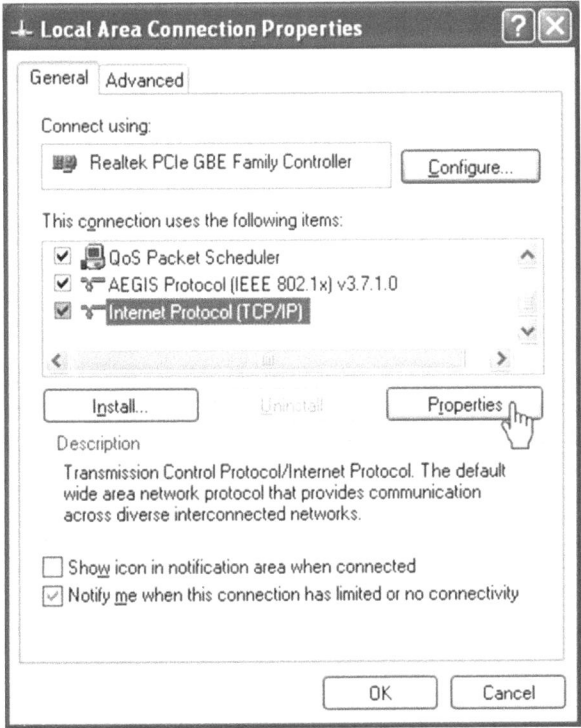

Fig. H.58 Windows XP Local Area Connection Properties (Vixen)

In the "General tab", mark check boxes for "Obtain IP address automatically (O)" and "Obtain DNS server address automatically (B)". Click the OK button (Fig. H.59).

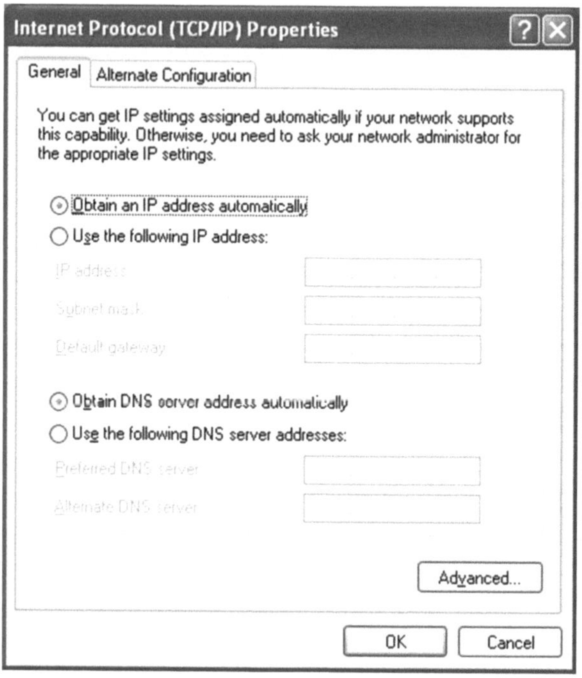

Fig. H.59 Windows XP Internet Protocol Properties (Vixen)

Go to "Control Panel" and select "Internet Option" in "Network and Internet" dialog box (Fig. H.60).

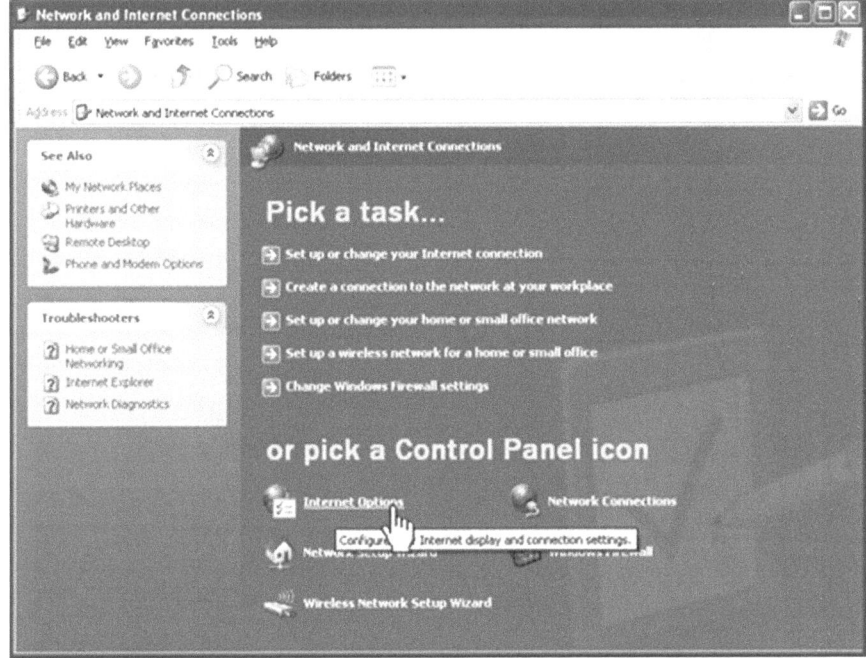

Fig. H.60 Windows XP Network and Internet Connections (Vixen)

"Internet Protocol Properties" are displayed. Go to "Connection" tab and select "LAN Settings" (Fig. H.61).

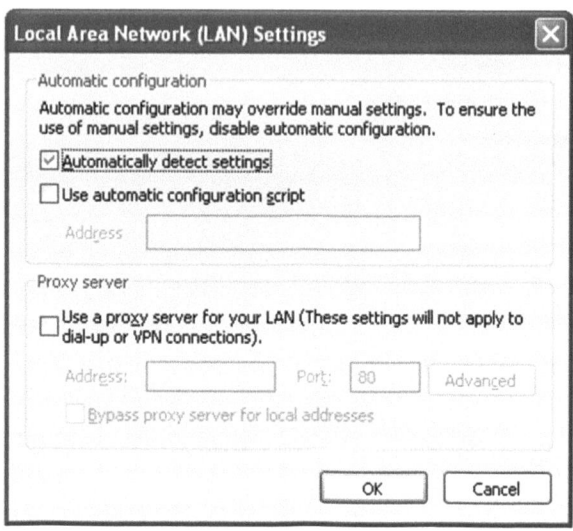

Fig. H.61 Windows XP Local Area Network Settings (Vixen)

Confirm that a check box for "Use Proxy Server for LAN" in the lower center is not marked. If the box is checked, remove the check mark. (Remember to return to the previous settings when you finish the download.)

Go back to the procedure on "How to Install the Comet Registration File".

Biographies

Author—James L. Chen: Retired Department of the Navy and Federal Aviation Administration Radar and Surveillance Systems Engineer. Guest lecturer at local Washington DC/Northern Virginia/Maryland astronomy clubs on amateur astronomy topics of eyepiece design, optical filters, urban and suburban astronomy, and lunar observing. Author of an Astronomy Magazine article on Dobsonian telescope design in November 1989 issue. First book, published in June 2014 by Springer, entitled *How to Find the Apollo Landings* Sites. Second book entitled *A Guide to the Hubble Space Telescope Objects,* is also available from Springer. Served as a part-time technical and sales consultant for two Washington DC area telescope stores for over 30 years.

Graphics Designer—Adam Chen: Former Program Manager of media support for NASA Headquarters in Washington DC. Creator and executive producer of major NASA publications, including the book and web-book application documenting the history of the Space Shuttle Program "Celebrating 30 Years of the Space Shuttle Program. Currently works in marketing for Brown Advisory, an investment firm in Baltimore, MD.

© Springer International Publishing Switzerland 2016   289
J.L. Chen, A. Chen, *The Vixen Star Book User Guide*, The Patrick Moore Practical Astronomy Series, DOI 10.1007/978-3-319-21593-8

Index

© Springer International Publishing Switzerland 2016   291
J.L. Chen, A. Chen, *The Vixen Star Book User Guide*, The Patrick Moore
Practical Astronomy Series, DOI 10.1007/978-3-319-21593-8